Recycle Based Organic Agriculture in a City

Seishu Tojo

Editor

Recycle Based Organic Agriculture in a City

 Springer

Editor
Seishu Tojo
Institute of Agriculture
Tokyo University of Agriculture and Technology
Fuchu, Tokyo, Japan

ISBN 978-981-32-9874-3   ISBN 978-981-32-9872-9   (eBook)
https://doi.org/10.1007/978-981-32-9872-9

This Springer imprint is published by the registered company Springer Nature Singapore Pte Ltd.
The registered company address is: 152 Beach Road, #21-01/04 Gateway East, Singapore 189721, Singapore

Preface

You should feel relieved when you see the spread farmland on the train from the city center going out to the suburbs because you realize that such farmland and farming activities there are tightly related to your essential daily life. Farmland is a source of life and provides important resources of our daily staples. We hope that the city dwellers who are living densely in a limited space can feel worthwhile to be closely connected with the diverse living things on the small farmland through urban agriculture.

Although urban agriculture has no concrete definition, we define it in this book as a type of new farming that is operated in and around urban regions. There are various categories of urban agriculture besides growing high-valued crops such as fruit, flowers, or ornamental plants in relatively small fields. The types of agriculture that intend to produce not solely ordinary agricultural products occupy an important place in urban agriculture. Harvesting crops through gardens by urban residents is becoming a way for them to enjoy farming and a new lifestyle. Being called plant factories, modern greenhouse horticulture that is independent on open-spaced farmland or soils may be operated in buildings using nutrient solutions; this type of farming has recently become popular as well. These various farming practices that have characteristics or functions uniquely suitable for city residents are expected to expand even more in the future.

Here in this book, however, we will discuss a type of urban agriculture that depends on open farm fields and soils. We also hope that the urban agriculture we will create includes as many advanced technologies and as much knowledge and wisdom as possible to cherish the connections of lives in the future. We wish to contribute to solving global environmental issues and simultaneously providing safe and natural foods to citizens that conventional farming technologies have not yet well-prepared especially for urban residents.

Many incidents such as food fraud have occurred recently in Japan, and every time we hear about these scandals, many of us long for safe agricultural products even more. Similarly, when the price of vegetables goes up because of frequent natural disasters around the country, we feel the need for a stable supply of less expensive agricultural products. On the other hand, possibly due to global warming, rice harvested in Hokkaido, which is located in north of Japan, becomes rich in taste as that grown in central Japan, and tropical fruits like passion fruits or dragon fruits can now be cultivated in central Japan such as Chiba Prefecture. The apparent change in the weather and agricultural environment concerns city residents in terms of how long one can sustain a stable food supply. Some people wish and feel safe that the produce cultivated within the city would supply the basic food requirements. Most urban residents only hope that the food must be fresh, safe, and nutritious. Our goal is to realize that urban agriculture will supply quality produce to urban residents.

A sustainable biochemical cycle is indispensable for the practice of organic agriculture. The vital force of life is strong, but we cannot maintain a steady supply of quality food without a sound stock and flow of nutrients and mineral components. Also, material balance is an essential indicator for managing crop productions while maintaining the global environment. What kind of ingenuity is required to establish a steady supply of crops in a city that is not necessarily suitable for crop production? This book provides partial solutions by presenting a showcase of our current research activities along with some recommendations to implement the findings, although some technologies still require further studies and social validations as well. Water is the most critical factor in agriculture; crops cannot grow without water. We do not, however, include the water cycle in this book. Nevertheless, we should consider improving the existing urban water supply system by integrating the agriculture-based water recycling and reuse facilities to realize a recycling-based agriculture practice in urban areas.

In this book, we also focus on effective recycle of wasted biomass, organic wastes generated in cities to achieve organic farming. Among the various methods to recycle organic waste, instead of using composting that has been used traditionally, carbonizing the organic waste for a low-carbon society has also become a focus of attention in recent years. Furthermore, in addition to producing recyclable materials, carbonization also converts organic waste into energy, and this idea would be in line with the philosophy of organic agriculture. Recycling the discarded organic matter in an appropriate manner and reusing it as both material and energy should be the basic concept of a new urban organic farming. It can be said that this practice is a big challenge to the next generation of urban agriculture for realizing a safe and organic agricultural production with only a small environmental load by positively incorporating various advanced research results.

My colleagues have contributed to writing manuscripts, but researchers in Thailand and Egypt also contributed significantly to this book, especially on the carbonization technology. In addition, Indonesian researchers have helped with chapters presenting the corporate organic farming practices in Indonesia. I would also like to thank all authors who have made contributions to the completion of this book. I wish to appreciate Dr. Mei Hann Lee of Springer's Editorial Department for suggesting and encouraging us from the start and Mr. Selvakumar Rajendran for helping with editing this publication. Finally, I would like to express my sincere appreciation to Dr. Allen C. Chao for reviewing all the manuscripts from the standpoint of a specialist and linguist.

Fuchu, Tokyo, Japan   Seishu Tojo
                                                                                  (on behalf of the authors)

Contents

Contributors

Hajime Araki Field Science Center for Northern Biosphere, Hokkaido University, Hokkaido, Japan

Miftahul Choiron Department of Agro-industrial Technology, University of Jember, Jember, Indonesia

Tadashi Chosa Institute of Agriculture, Tokyo University of Agriculture and Technology, Fuchu, Tokyo, Japan

Haytham M. El Sharkawi Agricultural Research Center (ARC), Giza, Egypt

Takahiro Ito Niigata Agro-Food University, Niigata, Japan

Hiroyuki Kashima Super-cutting-edge Grand and Advanced Research (SUGAR) Program, Japan Agency for Marine-Earth Science and Technology (JAMSTEC), Yokosuka, Japan

Hitoshi Kato Division of Lowland Farming, Central Region Agricultural Research Center, National Agriculture and Food Research Organization, Tsukuba, Japan

Rei Kikuchi Division of Farming Systems Research, Western Region Agricultural Research Center, National Agriculture and Food Research Organization, Fukuyama, Japan

Masakazu Komatsuzaki College of Agriculture, Ibaraki University, Ami, Japan

Vicheka Lorn United Graduate School of Agricultural Science, Tokyo University of Agriculture and Technology, Fuchu, Tokyo, Japan

Tineke Mandang Bogor Agricultural University, Bogor, Indonesia

Pisit Maneechot Excellent Center for Asia Pacific Smart Grid Technology (APST), Naresuan University, Phitsanulok, Thailand

Takashi Motobayashi Field Science Center, Tokyo University of Agriculture and Technology, Fuchu, Tokyo, Japan

Toru Nakajima Department of Bioproduction and Environment Engineering, Tokyo University of Agriculture, Setagaya-ku, Tokyo, Japan

Yosei Oikawa Institute of Agriculture, Tokyo University of Agriculture and Technology, Fuchu, Tokyo, Japan

Y. P. Sudaryanto Bina Sarana Bakti (BSB) Foundation, Bogor, Indonesia

J. Indro Surono Bina Sarana Bakti (BSB) Foundation, Bogor, Indonesia

Prapita Thanarak School of Renewable Energy and Smart Grid Technology (SGtech), Naresuan University, Phitsanulok, Thailand

Seishu Tojo Institute of Agriculture, Tokyo University of Agriculture and Technology, Fuchu, Tokyo, Japan

Megumi Ueda United Graduate School of Agricultural Science, Tokyo University of Agriculture and Technology, Fuchu, Tokyo, Japan

Dian Askhabul Yamin Bina Sarana Bakti (BSB) Foundation, Bogor, Indonesia

Tiejun Zhao Niigata Agro-Food University, Niigata, Japan

Chapter 1
Status and Prospects of Urban Agriculture

Seishu Tojo

Abstract Cities in Japan have farming fields reserved with the Productive Greenery Land Law enacted in 1974. This act exempts the farmland owner in an urbanization-designated area from paying the land tax equivalent to that of a home lot by carrying out the farming practice on their property. Challenges faced by farmers in urban farming include high taxes such as inheritance and real estate taxes, the environment less fitted for farming operations, aging and shortage of farming successors, and difficulty in management expansion.

Farmlands in urban regions with harsh conditions of soil such as high pH level and low exchangeable lime content management are growing in number. The continuous accumulation of some specific nutrients by application of chemical fertilizer causes unbalanced soil nutrient. Soil organic matter indicates whether the soil condition is in a suitable state for plant to root or microorganisms to grow so that it plays a vital role in providing plants with nutrients, water, and other components. The recent global warming makes it necessary to measure soil carbon storage capacity and dynamics. Researchers suggest that soil organic matter will increase through continuous organic farming. For a sustainable organic farming practice, the increase of soil organic matter is essential and eventually contributes to alleviating global warming by fixing carbon dioxide in the soil.

The importance of constructing a recycling system that utilizes many kinds of urban organic resources such as wasted biomass and creating participatory agriculture with the support of citizens will be stated for realizing sustainable agricultural management in urban region. Recycling organic resources as crop nutrient and renewable energy source from wasted biomass using new technologies is also desired to establish urban organic agriculture.

S. Tojo (✉)
Institute of Agriculture, Tokyo University of Agriculture and Technology, Fuchu, Tokyo, Japan
e-mail: tojo@cc.tuat.ac.jp

© Springer Nature Singapore Pte Ltd. 2020
S. Tojo (ed.), *Recycle Based Organic Agriculture in a City*,
https://doi.org/10.1007/978-981-32-9872-9_1

1.1 Legal Standpoint of Urban Agriculture in Japan

Japan has seen a significant expansion of urbanization owing to rapid economic growth after World War II. In 1968, the government enacted the City Planning Act to draw a line between designated and controlled areas of urbanization. The urbanization-designated area consists of 3.9% of Japanese soil, where urbanization should proceed preferentially in a well-planned manner. An urbanization-controlled area, on the other hand, is the area where urbanization must be controlled and, in principle, improvement of urban facility should not be conducted. The urbanization-controlled area consists of 10.3% of Japanese soil.

The government also enacted the Productive Greenery Land Law in 1974 that is the act concerning agricultural land among others reserved in the urbanization promotion area for developing an excellent urban environment in harmony with agriculture, forestry, and fishery. The Act has preserved farming fields in the urbanization-designated area. An amendment to this Act in 1992 exempted the owner of farmland in an urbanization-designated area from paying the land tax more than the amount equivalent to that of a home lot. The exemption is offered in exchange for landowners to carry out green farming operation practice on their property for 30 years. There are several criteria for the property to be recognized as a productive green area; the agriculture, forestry, and fishery productions must be sustainably practiced on the productive green land with area of more than 500 square meters. Once a lot is recognized as productive green land, the owner cannot subject their property to other purposes except for farming. They can also neither build, rebuild, or expand buildings in the subject area nor utilize the area for different uses such as residential land development. Because this Act will expire in 2022, revision of the Act has been discussed lately, and the amendment in 2017 has mitigated several regulation criteria such as the minimum requirement on acreage and limitations on the use or practices of the property. This amendment focuses on establishing the system for designating specific productive green areas. It enables the landowners of relatively small farmlands to apply their property as a productive green land, to run stores selling crop products directly to consumers, or to operate farmland restaurants as allowed by the amendment. Currently, the landowners are allowed to extend their productive green designation period by 10 years upon request.

Until the recent declining and aging farming population, the farmland in urban areas has been the target of residential development; the urban farmland has gradually shifted to a more valuable asset instead of a source to only supply fresh crops to local consumers. The land is also valued by residents as a disaster-preventing space, a fine scenery spot to preserve the environment, or a place to provide a chance for the community to interact socially with farming experience. Also, the "Basic Law on Promoting Urban Agriculture" enacted in 2015 that clearly defines the effective utilization and proper preservation of farmland in urban areas widely recognizes the significance of urban farming and agriculture.

The basic law defines the need for favorable coexistence of urbanization with agriculture and obliges the government and local governments to formulate action plans for promoting urban agriculture. Specifically, the law includes the following provisions: (1) improving crop supply flow and training and securing suppliers; (2) creating disaster-preventing spaces and fine scenery, fulfilling the preservation mechanism of national land and the environment; (3) developing an action plan to formulate a precise land use; (4) taxation measures on the land in urban areas that continue its use for agriculture; (5) promoting consumption of local crop products in the local area; (6) improving the community environment by providing farming experience; (7) providing students with farming experience in their school education; (8) enhancing awareness of the nation about agriculture; (9) promoting learning of farming techniques/knowledge for city dwellers; and (10) promoting research and study on the field.

1.2 Agricultural Management in City

1.2.1 Status of Agricultural Management

According to the questionnaire survey conducted in 2011 by the Ministry of Agriculture, Forestry, and Fisheries of Japan (MAFF of Japan 2011) and in 2015 by the Census of Agriculture and Forestry, farmland in the urbanization-designated areas consists of only 2% of all farmland in Japan, or 72,000 ha out of 4.471 million ha. Regarding the sales amount, the farmland produces crop worth 446 billion yen that is 8% of all crop product sales of 5836 billion yen. Although the average scale of 0.75 ha farmland by urban farmers is relatively small, urban farmland is advantageous being near the consumers so that various crops mainly fresh vegetables can be supplied to the customers. The sales by 60% of urban farmers amounts to 1 million yen per year, and the average sales amount is 2.7 million yen. The primary source of livelihood for most urban famers comes from the real estate business with 10% of the sales exceeding 7 million yen per year. Urban crops consist of 69.3% of vegetables grown outdoors, 60% of paddy rice, 32.6% of potatoes and beans, 18.25% of fruits, 12.1% of greenhouse crops, and 8.4% of flowers and ornamental plants. There are no sales of animal production in cities. The gross rate of crops output in urban region is as shown in Fig. 1.1. As the destination of their sales, 41.9% go directly to consumers, 41% to the Japan Agricultural Cooperatives (JA), 30.2% to the wholesale market, 22.4% to the farm stand, and 8.3% to supermarkets.

Challenges faced by farmers in sustainable urban farming revealed in available reports include high taxes such as inheritance and real estate taxes, low crop sales price, the environment less fitted for farming operations, complaints from surrounding residents, aging and shortage of farming successors, and difficulty in management expansion, among the many others. These issues are addressed by agricultural

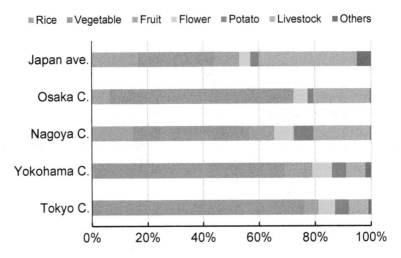

Fig. 1.1 Gross rate of agricultural crop production in big cities of Japan
Source: Municipal gross of agricultural crop production (2015), Statistics of Japan

promotion policies such as improving farming machinery, providing irrigation channels and drainage, training potential farming successors, and improving community farming practices or greenhouses.

1.2.2 Agriculture in Koganei: A Tokyo Bedroom Town

Koganei is a city located almost in the center of Tokyo, about 25 km from the metropolitan area; it is 4.1 km in width from east to west and 4.0 km in length from north to south with a total area of 11.33 km². The population is about 110,000, and most residents work in Tokyo. The JR (Japan Railway) Chuo Line runs through the city center from east to west, whereas Nogawa River runs in the south and Tamagawa Aqueduct in the north. Once there were rice paddies around Nogawa shed, but by 1960 the rice paddies were all abolished because of deteriorating water quality.

According to the Promoting Plan of Agriculture in Koganei, all farming areas in Koganei, which are located in the urbanization-designated area, have been reduced by 9.4% in 5 years from 89 ha in 2005 to 81 ha in 2010, and they are scattered among residential districts (Fig. 1.2). The reduction rate of areas with productive greens, on the other hand, was controlled at 8.2% in 5 years from 73.8 ha in 2005 to 67.7 ha in 2010.

The number of farming households was 346 in 1970, but it reduced by half in 40 years to 169 in 2010. About 30% of the farming-related business entities in the area, or 32 houses, are full-time farmers; 40%, or 41 houses, are semi-full-time farmers; and 30%, or 39 houses, are part-time farmers. The net sales amount to about 300 million yen, which shows slight increasing trend. Spinach, potato, and

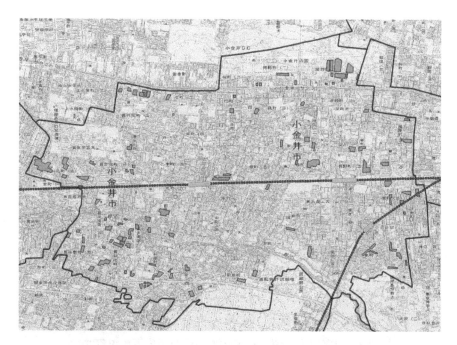

Fig. 1.2 Farmland in Koganei city, Tokyo. Green color mark indicates farmland
Source: Geospatial Information Authority of Japan

radish are the most popular vegetables cultivated in 3-ha areas. As for fruits, chestnuts cultivated in 10-ha areas are the most popular, followed by prunes and Japanese pears and then flowers and ornamental plants with 1-ha cultivation. There are 67 on-site sales shops, 1 cooperative store, and 4 antennae shops in the city. There are also four community farms, two farms for elderly citizens, two community farms opened to public for farming experience, and one community farm operated by farmers.

Poll of urban residents on their viewpoint on the role of urban farming shows the highest of 384 votes for "agricultural crop supply" followed by 237 for "preservation and formation of the natural environment," 179 for "place for work and living," and 173 for "educational or recreational opportunities" (Fig. 1.3). The result suggests that the role of urban farming is for not only growing the crop but also enriching the daily living of local residents (Koganei city 2011). Some farmers attempt to use composts made from kitchen waste as soil amendments for safe production of vegetables. These recycling activities are promoted by the city government for attaining both objectives of reducing discharged kitchen wastes and enhancing farming worth in city. The amount of reduced kitchen waste was 55 tons/year, and produced compost from kitchen waste was 58 tons/year in 2012. Farmers who use such composts can sell their agricultural products labeled with the use of compost from kitchen waste in Koganei city to local residents at JA shops and supermarkets. They have another sale event to exchange their products with a coupon that local

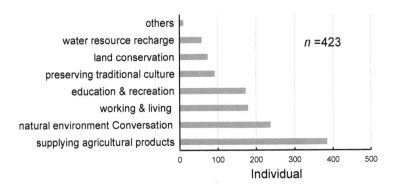

Fig. 1.3 Role of agriculture from the city dwellers' viewpoint in Koganei city
Source: Promoting Plan of Agriculture in Koganei (2011)

residents get from the city government for providing kitchen wastes to manufacture composts. This trend of activities will sustain in a city that farmers and residents have high safety mind of food and environmental consciousness.

1.2.3 Expectation of Urban Agriculture and Management Innovations

In 2011, the government of Japan enacted the "Sixth Industry Promotion Act" for creating a new add-on value by producing agricultural products in rural areas. The objective is to diversify and sophisticate traditional agriculture to enable the agricultural practice to extend beyond traditional practice; farmers can provide meals cooked from vegetables they harvest on their farms at farmers' restaurants in the urban area. It enables the actual practice of so-called agribusiness, and farming-oriented business management, as well as opens a way to obtain stable incomes for urban farmers from the farmland they succeed (Oda et al. 2014).

Aging of farmers is a more devastating problem that requires immediate attention. More than 50% of active farmers are over 65 years old, and one-third of the farms do not have successors that will significantly affect the succession of farmlands. Promoting agribusiness with a stable income by providing a steady yearly business environment is the key to preserving farmland in urban areas. Contemporary agriculture business management, which is expected to fulfill social responsibilities as described in Good Agricultural Practices (GAP), includes securing food safety, preserving the environment, securing labor safety, and maintaining animal welfare, among others. The trend is to shift agriculture management to a whole different practice as compared with traditional farming practices. Risk and information management that is essential to dealing with various risks including food contamination should be practiced along with information management by collecting, sorting, and delivering information related to crop production and sales operations (Nanseki 2012).

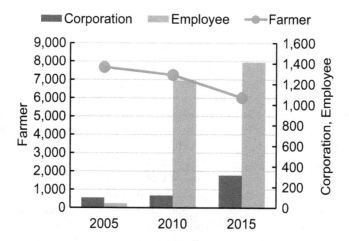

Fig. 1.4 Change in number of farmers and employees of agricultural corporation in Tokyo Source: Agricultural management entity, agricultural labor. Census of agriculture and forestry. Statistics of Japan

The traditional farming practice in Japan was small family management, and the management practice that was passed on from generation to generation is the biggest challenge. The aging issue of farmers has led to a gradual expansion of the farmland area managed by each farm and eventually to a shift of the labor force from a family-oriented to outsourcing practices (Fig. 1.4). As for urban agriculture management, most of which is probably family-oriented for the time being; the same change in employment-oriented family management or complimentary family management agribusiness should be inevitable (Matsuki 2011).

1.2.4 Attempts in World Cities for Stable Food Supply and High Self-Sufficiency

The year 2008 is said to be remarkable in human history with the urban population outnumbering the rural population for the first time. Similarly, 53% of the total population in Japan live in cities with population size over 200,000, and others live in smaller cities, towns, and villages so that 91% of the total population is concentrated in urban regions. Indeed, the situation requires stable and self-sufficient supply of food to cities. The days of free trade of agricultural produce will soon be over, and food supply will be a critical operation so that city dwellers will have to be self-sufficient. Innovation of the food production system still has a long way to go; those who live in cities are now beginning to consider how to sustain food for themselves against the inevitable shortage of oil, water, and land (Cockrall-King 2012).

In Paris, France, urban agriculture has been widely practiced since the nineteenth century, when the carriage was the primary means of transportation with abundant horse dung to be used as manure. There were 8500 farmhouses in Paris, and 100,000 tons of produce are harvested per year from the 1400-ha farmland. At the end of the nineteenth century, however, automobiles began to replace the carriage that led to a shortage of horse dung and eventually the rapid decline of urban agriculture. Many of the farmhouses in Paris were said to have quit farming because transportation of crop products for a long distance by automobiles becomes economically feasible. The number of farms in Paris was reduced to 800 houses by 1997. Lately, a civilian movement called "Green Hand" has become popular that brings farms back to the area by utilizing vacant lots for crop cultivation (Cockrall-King 2012).

Many city administrators worldwide recognize the importance of urban agriculture. For example, Havana, the capital of Cuba, is famous for urban agriculture in the 1990s. Artificial gardens known as "Organoponico" with an area of 0.2–0.5 ha were constructed by using concrete and soil supply 5% of agricultural food in Havana (Yoshida 2008). In Russia, as a result of the food crisis hit by the collapse of the Soviet Union, urban dwellers started cultivating idle lands and spaces. In Detroit (USA), the steel industry, which was once the key industry in the United States, is now declining. The local residents are focusing their attention on urban agriculture utilizing vacant lots available from closed mills (Atkinson 2012). In the United Kingdom, after the 2012 London Olympics/Paralympics games held in London, some people are beginning to build a new, farmland-oriented city in cooperation with public and private sectors by reforming various types of regions into farmlands (Martin et al. 2016). The largest city of Asia, Jakarta, Indonesia, recognizes that urban agriculture is an essential policy of city administration by creating urban agriculture in which many residents can positively participate including not only farmers but also women or younger people. Even in a developed country like the United States, where the overall food supply is more than enough, some urban areas, such as the so-called food desert or food refugees, face shortage in food supply. Cities in the world are now looking for a new form of society or lifestyle.

1.3   Change in Soil of Urban Farmland

Abandoned farmland in mountainous areas of Japan has become an issue because lately it has increased gradually in suburban regions during the progress of urbanization. Farmers in suburban areas allocate their labor force to other lines of occupations other than agricultural activities and depend on the income they earn from these nonagricultural occupations. As a result, lack of sufficient labor force caused the farmland to be abandoned (Hattori and Yamaji 1988). Restoring abandoned farmland is not an easy task so that these abandoned lands are hardly sought after by investors or farmers. One abandoned field will also affect the adjacent fields; uncontrolled weeds grow wildly in the abandoned area and scatter their seeds into adjacent fields, or wild animals wander and damage the fields nearby thus making it hard to maintain proper cultivation conditions for the neighboring field. Also, the situation makes it problematic to maintain proper working conditions for the infrastructures shared with the neighboring community such as farm roads and irrigation canals; this affects not only the condition of the field but also production structures relating to farming.

Available reports indicate that community farms in urban regions that experience harsh conditions of soil management are growing in number. Anzai et al. (2012) reported some of the harsh soil conditions such as high pH level and low exchangeable lime content. The continuous accumulation of some specific nutrients causes unbalanced soil nutrient. The report also pointed out that the community farm textbook or brochure encourages the citizen farmer to apply garden lime and calcium once every year without checking excessive accumulation of specific nutrient. Yoshida et al. (2008) found that soil samples collected at 48 sections of farmland in Tokyo contained excessive phosphate concentrations with an average 95 mg/100 g soil of available phosphate and 18 mg/100 g soil of soluble phosphate. This finding shows that agricultural production using urban farmland requires an adequate policy to achieve sustainable soil management by practicing a precise soil measurement technique.

The organic matter contained in the soil is often used as an indicator to measure the land fertility. Soil organic matter indicates whether the soil condition is in a suitable state for plant to root or microorganisms to grow so that it plays a vital role in providing plants with nutrient, water, and other components. The recent global warming makes it necessary to measure soil carbon storage capacity and dynamics; soil organic matter is a critical indicator in that regard. Komatsuzaki and Faiz Syuaib (2009) reported that the level of soil carbon in western Java, Indonesia, reads 2.31% in the rice paddies of conventional farming and 3.24% in the paddies with a 4-year practice of organic farming. The results suggest that soil organic matter would increase through continuous organic farming. Thus, for a sustainable organic farming practice, the increase of soil organic matter is essential and eventually contributes to alleviating global warming by fixing carbon dioxide in the soil (Lebel et al. 2007).

1.4 Prospect of Organic Agriculture in City

Agricultural manufacturers including crop-producing farmers and crop-processing producers must pass organic certification in order to be licensed to sell their agricultural products under the label "Organic" in Japan (Tojo et al. 2008). The yearly review of the licensing by the authorizing organization is required for the manufacturers to be licensed continuously. "Organic" is defined based on the 1999 amendment of Japanese Agricultural Standards (JAS) in accordance with the Food and Agriculture Organization (FAO) codex guideline regarding agricultural products and their processed products. What is defined as "Organic JAS" is crop production management based on enhancement of the preservation of the natural recycling mechanism in agriculture. This is carried out by avoiding the use of chemically synthesized fertilizers or pesticides to alleviate environmental loads caused by practicing organic farming as much as possible. Specifically, it defines that healthy soil prepared with manure must be practiced and the use of chemically synthesized fertilizers or pesticides is generally prohibited for more than 2 years for annual crops and 3 years for perennial crops prior to seeding or planting, as well as during the period of cultivation. Furthermore, the use of transgenic seed and seedling is also prohibited along with detailed regulations on the following items:

1. Location of the field, farm, and cultivation ground
2. Type of crop production
3. Cultivation area specifications
4. Dates and contents of work
5. Specifications of seed or seedling used (including seedling, sapling, graft, stock, or any other part of the plant organism used for propagation, excluding seeds) or names and the amount of use or purchase of inoculum
6. Names and amount of agricultural chemical materials used
7. Names and management policy of machinery and equipment used
8. Management policy of postharvest, including harvesting, reception, transportation, sorting, preparation, washing, storing, and packaging, among the many others

Regarding the import/export of agricultural products, only certified products from countries that have approved organic programs equivalent to those implemented in Japan shall be considered as "organic products." As of March 2018, nations with approved organic equivalence with Japan were the EU (28 countries), Australia, the United States, Switzerland, Argentina, New Zealand, and Canada.

This strict certificate system has sorted out the once chaotic situation regarding domestic organic agricultural products, but sales of these products are still on a limited scale. There were about 70 domestic or foreign organizations registered as certification authorities in 2017, and 3678 farms were certified as organic crop producers. A total of 10,366 ha that has been approved as organic farmland consists of only 0.23% of all farmland in Japan. The sales volume, mostly in vegetables, is gradually increasing, but it is only 60,000 tons per year (Fig. 1.5). The produce prices vary according to the crop type; organic vegetables cost 1.3–1.8 times more than the price of noncertified organic vegetables (Fig. 1.6).

Organic farming abroad is changing drastically. In the United States, the practice took root in the 1960s. Since then, organic farming has been supported by small-scale farming, but now the practice has been gradually industrialized. By contrast, agro-industries are entering the market of organic farming to bring about a huge change. The US Organic Foods Production Act (OFPA) enacted in 1990 clearly defines the practice, attitude, and philosophy about organic farming as the approach based on the material loaded in the practice. Later, in 2000, OFPA began to function according to the National List of Allowed and Prohibited Substances enacted that year. The definition of organic farming by prohibiting the use of chemical materials in an agroecological and community-based production process has made the US food market favor the large-scale organic agricultural business and industrial enti-

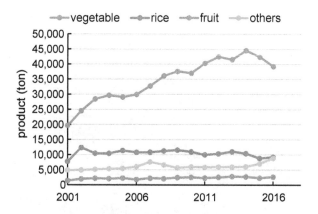

Fig. 1.5 Organic agricultural products granted by certifying authority in Japan
Source: Track record of organic agricultural products (2001–2017). MAFF of Japan

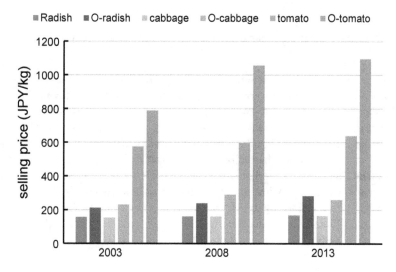

Fig. 1.6 Change in selling price of organic vegetables in Japan
O-radish, O-cabbage, and O-tomato mean organic vegetables
Source: Trend survey of fresh vegetables (2013). Statistics of Japan

ties because they satisfy the market demand of industrialized food. The sales amount of the organic industry was 3600 million dollars in 1997, increased to 7800 million dollars in 2000 when organic vegetables began to be sold in supermarkets, and further skyrocketed to a 39-billion-dollar market in 2015 when sales of organic produces became regular in supermarkets (Fitzmaurice and Gareau 2016). The production and sales style of this organic supermarket has indeed changed the essential meaning of "organic" and symbolized the transformation of organic farming into agri-business.

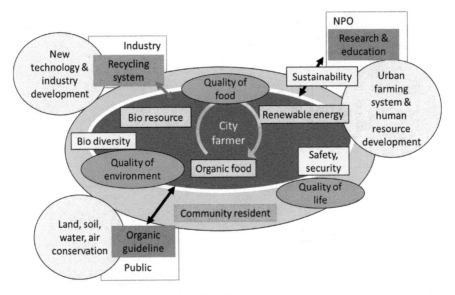

Fig. 1.7 Fundamental structure of recycle-based organic agriculture in a city

Cooperating with small-scale family farmers by organizing an urban-type agricultural management body warrants channels for obtaining necessary materials and selling agricultural produces. This is necessary in order to establish organic farming in the city formed seminaturally and provide safe and reliable food to citizens while preserving valuable small farmland and surrounding environment. Sustainable agricultural management will be realized by constructing a recycling system that utilizes many kinds of urban organic resources such as wasted biomass and creating participatory agriculture with the support of citizens as shown in Fig. 1.7. Enhancing researches on recycling organic resources as crop nutrient and renewable energy from wasted biomass and education on agriculture using new technologies is also desired to establish urban organic agriculture.

References

Anzai T, Hidaka T, Yahagi M, Kato T, Inagaki T (2012) Reality of soil cultivating vegetables in allotment garden. Agric Hortic 87(12):1171–1178

Atkinson AE (2012) Promoting health and development in Detroit through gardens and urban agriculture. Health Aff 31(12). https://www.healthaffairs.org/. http://dx.doi/10.1377/hlthaff.2012.1106. Accessed 12 Jan 2019

Cockrall-King J (2012) Food and the city: urban agriculture and the new food revolution. Prometheus Books Inc, Amherst

Fitzmaurice CJ, Gareau BJ (2016) Organic futures: struggling for sustainability on the small farm. Yale University Press. (Japanese translation) Sekine K, et al. (2018) Tsukuba-shobo, Tokyo

Hattori T, Yamaji E (1988) Factors to abandon cultivation by the condition of farm household in suburban area. J Rural Plan 16(4):325–333

Koganei City (2011) Promoting plan of agriculture in Koganei. https://www.city.koganei.lg.jp/
kurashi/nogyo/...sakutei.../nougyousinkoukeikaku.pdf. Accessed 14 July 2018
Komatsuzaki M, Faiz Syuaib M (2009) A case study of organic rice production system and soil
carbon storage in West Java, Indonesia. Jpn J Farm Work Res 44(3):173–179
Lebel L, Garden P, Banaticla MRN, Lasco RD, Contreras A, Mitra AP, Sharma C, Nguyen HT, Ooi
GL, Sari A (2007) Integrating carbon management into the development strategies of urbanizing regions in Asia – implications of urban function, form and role. J Ind Ecol 11(2):61–81
MAFF of Japan (2011) Report of fact-finding investigation on urban agriculture in Japan. http://
www.maff.go.jp/j/nousin/kouryu/tosi_nougyo/pdf/tosi_tyousa_honntai.pdf. Accessed 14 July
2018
Martin G, Clift R, Christie I (2016) Urban cultivation and its contributions to sustainability: nib-
bles of food but oodles of social capital. Sustainability 8(5):409
Matsuki Y (2011) The logic and state concerning new types of business organization of family
agricultural holdings – a comparative analysis between Japan and western countries. Bull
Nippon Vet Life Sci Univ 60:79–98
Nanseki T (2012) Food risk and farms in the next generation: issues and perspective.
Nogyokeizaikenkyu 84(2):95–111
Oda S, Chomei Y, Kawasaki N, Nagatani T (2014) Outlook for innovative research on agricultural
networks –from the viewpoint of governance and conflict. Kyotodaigaku-Shigenkeizaikenkyu
19:73–94 (in Japanese)
Tojo S, Watanabe K, Kimura M (2008) Environmental load of Japanese dairy farm in life cycle of
milk production. CIGR international conference of agricultural engineering. Iguassu, Brazil
Yoshida T (2008) Urban agriculture in Cuba: The current situation and perspectives in the land
management system and political Initiative. Nouson-kenkyu 106:109–118. http://nodaiweb.
university.jp/noukei/pdf/NSO106_10.pdf. Accessed 20 Aug 2018
Yoshida A, Inagaki K, Kato Y, Goto I (2008) Soils of allotment garden and citizens experience
farm in Tokyo. Proc Jpn Soc Soil Sci Plant Nutr 54:272

Chapter 2
Discharge and Recycling of Urban Wasted Biomass

Seishu Tojo

Abstract Various types of organic waste are generated in urban areas. These organic wastes as recyclable biomass should be utilized for multipurpose resource such as fertilizer and fuel, among others. Food waste can be a target biomass because increasing awareness of the issue on food wasting in recent years has led to a decreasing trend in the quantity of food waste collected. The wasted biomass is also expected as an energy resource because it contains lots of carbon component. Pruned branches from bushes and trees along streets or in parks with a relatively low moisture content are desirable combustion materials. Waste cooking oil is specifically feasible to be converted into biodiesel fuel.

The compost consists of not only inorganic nitrogen, phosphorus, organic acid, or sugars from the decomposition of organic matter by microorganisms but also undecomposed residue, as well as dead and living microorganisms. It is generally mixed with soil and is intended to be used as a soil conditioner. The definition of compost is somewhat vague and complicated because there is no definite method applicable for qualifying and assessing compost. Compost is intended to function primarily as a soil conditioner, followed by acting as a nutrient provider. Raw materials for compost, aside from livestock excreta, include agricultural residue, fallen leaves, food waste, sewage sludge, etc. The materials vary so widely that each compost has unique characteristics according to the raw and secondary materials used.

Knowledge and information concerning the impact of recycled products on the processes of decomposition, mineralization, and crop intake, as well as their remaining residues in soil, will be the most fundamental concerns for promoting the practice of organic farming. An attempt using self-organizing map (SOM) is explained so that both manufacturers and consumers can understand various properties of recycled products.

S. Tojo (✉)
Institute of Agriculture, Tokyo University of Agriculture and Technology, Fuchu, Tokyo, Japan
e-mail: tojo@cc.tuat.ac.jp

© Springer Nature Singapore Pte Ltd. 2020
S. Tojo (ed.), *Recycle Based Organic Agriculture in a City*,
https://doi.org/10.1007/978-981-32-9872-9_2

2.1   Wasted Biomass in Urban Areas

2.1.1   Wasted Biomass Collected by Local Governments

Various types of organic waste are generated in urban areas and some of which cannot be clearly distinguished from municipal solid wastes. Collecting sorted garbage and charging fees for the service in urban areas is encouraged in Japan. The establishment of a collection system in which residents sort reusable resources from garbage is underway, and the system is gradually taking shape. Local governments are in charge of the collection and disposal of general household waste discarded by residents. However, the rules and regulations for sorting and collecting garbage vary among different local governments. For example, some local governments collect branches pruned from garden trees as combustible garbage, whereas some others classify these branches as organic resources. Likewise, some governments collect food scraps from homes as organic resources, and some other governments collect, process, and compost organic waste into organic fertilizer for local farmers. Author and colleagues recognize these organic wastes as wasted biomass which should be utilized for multipurpose resource such as fertilizer and fuel, among others.

Table 2.1 shows the typical wasted biomass discarded in urban areas. Several public organizations have recommended methods for analyzing the quantity and quality of urban wastes. For example, the List of Basic Data for Grasping the Biomass Amount Present in a Region (NARO 2010) published by the National Agriculture and Food Research Organization (NARO) is one such reference. Table 2.1 shows the output rate and composition of several typical urban wasted biomass. The amount of food scraps discarded is 217 g per person per day for primary residential units, and this waste contains 90% wet basis (w.b.) moisture with 44% carbon content and 1.4% nitrogen content. The total annual dry weight of food scraps collected in a city can be calculated by multiplying the value of the primary residential units by the number of residents in the area. The resulting value, of course, is just an indicator, and the actual amount will vary in each city depending on factors such as the geographical locations or diet of the residents. If the waste is

Table 2.1 Output rate and contents of organic waste in city and its environs

Organic waste	Output rate	Unit	Moisture content (% w.b.)	Carbon content (% DM)	Nitrogen content (% DM)
Kitchen waste	217	g/person/day	90	44	1.4
School lunch waste	0.06	kg/person/day	75	46	3.1
Park trimmed branch	1.8	t/ha/year	57	52	0.39
Sewage sludge[a]	0.55	t/person/year	97	37	9.2

Source: NARO, local energy unit, biomass data (2010), partially modified by author
[a]Sewage sludge is categorized in the industrial waste in Japan

disposed of by using incineration, the moisture content of the organic waste is an indicator of whether the target waste material will be self-sufficient in the combustion process without the addition of auxiliary fuel. Sano (2007) reported that the threshold for the moisture content of organic waste was 60% w.b. or less and preferably less than 50% w.b. for stable self-sufficient combustion. Therefore, most of the organic waste collected in urban areas will require auxiliary fuels in the combustion process.

2.1.2 Recycle of Wasted Biomass from Food Scrap

In general, there are three types of food waste: food scraps from homes, food wastes from small- or medium-size business offices, and waste collected by industrial waste disposal facilities from food manufacturing industries. Increasing awareness of the issue on food wasting in recent years has led to a decreasing trend in the quantity of food waste collected. Table 2.2 summarizes the recycled amount of different organic waste categories from various sectors of food industries such as food manufacturing, food wholesale, food retail, and catering. According to the "Act concerning the Promotion of Utilization of Recyclable Food Waste" (MAFF), each industrial sector is obligated to recycle the discarded food waste. The MAFF report shows that 71% of all food waste is recycled for reuse as organic materials; whereas 3% is recycled in thermal waste-to-energy (WtE) processes as thermal resources. Of the recycled materials, 18% are used as fertilizer including 46% of the waste discarded from the wholesale industry, 34% from catering service, and 32% from retailers. The mixture of various food wastes and fertilizers is the most popular targeted material for recycle.

Some of the food wastes from the catering industry contains seasoning agents, which makes the recycling a challenging process because they contribute significant salinity and either higher or lower pH to the wastes; these wastes are particularly unfavorable to be treated in a microbial fermentation process. Hence, there could be a large amount of organic waste that is not necessarily fit for composting or methane fermentation procedures.

Table 2.2 Food industrial waste and recycle in Japan (unit: 1000 t)

Recycle sector	Total	Fertilizer	Feed	Methane	Others	Heat
All food industry	19700	2512	10269	616	588	537
Manufacturer	16167	2205	9965	572	348	533
Wholesale	267	58	38	8	23	3
Retail	1271	152	205	30	95	1
Restaurant	1994	97	61	6	122	–

Source: MAFF of Japan (2016)

2.1.3   Energy Use of Wasted Biomass

2.1.3.1   Method of Energy Use

Carbon component contained in biomass is expected as an energy resource. The carbon content of the available wasted biomass in Japan is summarized in Fig. 2.1. In the city and its environs, pruned branches from bushes and trees along streets or in parks with a relatively low moisture content are desirable combustion materials. Yamagishi and Kurihara (2012), however, reported that the average moisture content of these branches was 50% w.b. with a higher heating value of 20 MJ/kg and lower heating value of 8 MJ/kg. Most of these branches in urban areas are incinerated to recover heat energy; but the incineration does not contribute to reducing CO_2 emissions.

Concerning the wasted biomass from crop residues, Motghare et al. (2016) reported that lower moisture in the residue results in higher heating value. For example, rice straw with a low moisture content of 8.6% dry basis (d.b.) has a higher heating value of 17.6 MJ/kg. However, crop residue often has a high ash content that is not fit for use as a combustion agent. Some promising reports indicate the possibility of using crop residue as the raw material to produce bioethanol so that further technological development in this area is expected.

The moisture content of either food scraps or sewage sludge is quite high, and the waste is mostly fermented to produce methane gas to be used as fuel. There are 42 installations with methane fermentation facilities in Japan to convert wastes such as food scraps from ordinary homes into methane to generate electricity (Ministry of the Environment website). The biogas power plant in Nagaoka, for instance, collects 65 tons of food scraps per day to generate 2.45 million kWh of electricity

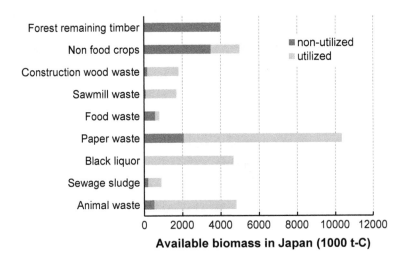

Fig. 2.1 Carbon content of available wasted biomass in Japan
Source: MAFF of Japan (2018)

annually, which is sufficient to supply electricity to 600 households. The construction cost was 1.9 billion yen, and plants of the similar scale are under construction in Kyoto and Kagoshima. Currently, there are 2150 sewage plants in Japan; 280 of these plants process sewage sludge by using methane fermentation, and 40 of these plants generate electrical power.

As for the types of food scraps, waste cooking oil is specifically feasible to be converted into biodiesel fuel (BDF). However, the high viscosity and low fluidity of the oil make it unfit as a liquid fuel. Hence, the oil is typically esterified using chemical agents such as methanol to be converted into fatty acid methyl ester for reducing its viscosity for fuel use. A relatively simple device would suffice for this procedure, but disposing of byproducts such as glycerol or cleaning waste liquid is problematic. Industrial edible oil waste is traded as a valuable commodity and used as feed additives, whereas residential edible oil waste collected by local governments is actively refined into BDF. In Kyoto, for example, 2000 collecting sites, each of which is equivalent to 300 households, are placed in the city to collect edible oil waste, and the refined BDF is used as the fuel to power city buses (B20 fuel, 20% of total fuel is BDF) or garbage collection vehicles (B100 fuel).

2.1.3.2 Feed-In Tariff Program of Biomass Electricity

The Feed-in Tariff (FIT) Program was introduced in Japan in 2012 to increase the use of renewable energy. In this program, the government warrants that electric power companies purchase the electricity generated using renewable energy sources at a certain guaranteed price for 20 years. Biomass is one of the target renewable energy sources in this program. The renewable energy surcharge, which is collected from users as a partial cost based on the amount of electricity used, was set at a unit price of 1.58 yen/kWh, whereas the price of normal home electricity is around 20 yen/kWh.

The purchase prices for biomass-derived electricity are listed in Table 2.3, and these prices differ based on the biomass source and power generation method. For instance, the price is 39 yen/kWh for electricity generated with methane gas produced from food waste, whereas the price drops to 17 yen/kWh for electricity generated from food waste combustion in incineration facility. At one time before FIT, the purchase price was 9 yen/kWh for electricity generated at methane fermentation facilities using livestock excreta. However, those days seem like very long ago as compared with the present situation regarding renewable energy. Moreover, the construction costs being amortized over a guaranteed purchase period of 20 years is an attractive condition.

Table 2.3 Feed-in tariff of renewable energy (biomass) of Japan (2015)

Biomass	Example	Price (yen per kWh excluding tax)	Year of procuration
Methane gas from methane fermentation	Material: sewage sludge, animal waste, food waste	39	20 years
Woody biomass from forest thinning (below 2000 kW)	Timber from forest thinning	40	
Woody biomass from forest thinning (over 2000 kW)		32	
General woody biomass, agricultural residues	Listing of lumber, imported timber, palm shell, rice husk, rice straw	24	
Construction waste	Wood construction waste (waste timber)	13	
Other wasted biomass	Pruned branch, woody waste, food waste, waste oil, sludge, animal waste, black liquid	17	

Resource: METI, Agency for Natural Resources and Energy

2.1.4 Composting of Wasted Biomass

2.1.4.1 Compost Manufacturing Method

The compost consists of not only inorganic nitrogen, phosphorus, organic acid, or sugars from the decomposition of organic matter by microorganisms but also undecomposed residue, as well as dead and living microorganisms. It is generally mixed with soil and is intended to be used as a soil conditioner. Various types of organic matter are used as the primary raw material for composting along with secondary materials that are used to condition the moisture level of the raw materials or ensure ventilation of the composting pile. The secondary materials consist of organic matter such as rice straw or sawdust, portions of the fermented compost used as the seeding material, and inorganic material such as pearlite. The proper standard moisture content of materials during the composting procedure is 60% on average, and the real moisture may differ according to the types of raw and secondary materials. The fermentation process is accomplished in two stages: the first fermentation stage can be completed in about a week using ventilated air to aerate with daily turnover of the content. Subsequently, the second fermentation stage takes 1–3 months for the compost to mature.

The "Bokashi" compost is a well-known and popular type of compost fermented by mixing soil with organic material. Mountain soil containing no pathogenic organisms is desirable although soil from fields or rice paddies may also be used. The first step involves mixing the soil with organic raw materials at a 50:50 volume ratio; the mixture is then piled up and left to ferment. The second step of the process is to turn over the content once every 1–2 weeks. The fermentation process will take 2–3 months to complete.

2.1.4.2 Restrictions on Compost Manufactured from Wasted Biomass

The compost manufactured from organic waste, especially from food waste, is categorized as a special fertilizer by the Fertilizers Regulation Act of Japan. The Act obligates manufacturers to report heavy metal component analysis results, but it does not require manufacturers to guarantee the content of the fertilizer nutrient components. The content of nitrogen, phosphorus, and potash in the compost accounts to 1–3% that is a small fraction of those contained in regular chemical fertilizer. The compost price is often set relatively low at approximately 5000–10,000 yen per ton.

The compost manufactured from sewage or human waste sludge is categorized as an ordinary fertilizer; manufacturers are obligated to indicate the fertilizer nutrient contents and report the results on heavy metal analyses. Table 2.4 lists the regulatory limitations on heavy metals; the values are so set that the amount of heavy metals accumulated in the soil should not exceed the tolerance value even if the compost fertilizer is used on the soil continuously for 100 years.

2.1.5 Use of Food Waste Compost in Community Farms

Kurita et al. (2010) reported on community farms that promote food waste recycling. Kitamoto, a city in Saitama Prefecture, has set up community farms since 1975 and food waste recycling farms since 1995. By 2010, the number of recycling farms increased to ten sites. Each section of these farms is 16.5 m² in size and can be used by residents for an annual fee of 1200 yen, which is almost half the price charged for the use of ordinary community farm plots. The total area of food waste

Table 2.4 Regulatory upper limit value of heavy metal

Heavy metal		Regulatory value
Arsenic	As	0.005%
Cadmium	Cd	0.0005%
Mercury	Hg	0.0002%
Nickel	Ni	0.03%
Chromium	Cr	0.05%
Lead	Pd	0.01%

Regulatory value is expressed by percent to dry matter weight
Source: Fertilizer Control Act of Japan

recycling farms in the city is 1.4 ha with 265 registered users, who are required to place fermentation bins in their designated section for collecting food waste to be composted in the bin. The fermented products are used as fertilizer, and this practice reduces the amount of food waste to be processed by the city. Most of the users are elderly people enjoying their retirement. The average useful life is 5.3 years, and the average distance from residential houses to the farm is 836.8 m. Kurita et al. noted that the community farms in Kitamoto have long been successful because their user organizations have promoted the sharing of farm management activities such as weeding, guidance on cultivation techniques, and exchange activities. It is a show-case of the community-oriented activities that began with the community farms, promoting exchange first between the users and owners of the farmland and then with surrounding residents. The success leads to the current practice of fermenting school-catering food waste for utilization in community farms.

2.2 Assessment of the Use of Wasted Biomass for Recycling

2.2.1 Measure of Compost Assessment

The definition of compost is somewhat vague and complicated, as mentioned previously, because there is no definite method applicable for qualifying and assessing compost. Generally speaking, compost is intended to function primarily as a soil conditioner, followed by acting as a nutrient provider. Raw materials for compost, aside from livestock excreta, include agricultural residue, fallen leaves, pruned branches, mowed lawn clippings, food waste, food scraps, and sewage sludge. The materials vary so widely that each compost has unique characteristics according to the raw and secondary materials used, thus making it difficult to have a uniform definition of the compost. Harada (2003) suggested an evaluation standard for assessing the maturity of compost by summing up the points assigned to individual conditions or processes, i.e., the color, texture, smell, moisture level, highest temperature during fermentation, fermentation period, number of stirring events, and absence/presence of enforced ventilation, among the many others, as summarized in Table 2.5. Recently, however, the range of biomass types used has widened because of the recycling trend; hence, this method based on external estimation of compost is no longer sufficient to fulfill this purpose.

2.2.2 Assessment Measure of Various Organic Materials

Recycling various organic wastes such as sewage sludge, food residue, or kitchen garbage has been promoted recently on top of recycling more conventional materials such as livestock excreta or food processing residues. In the recycling processes

Table 2.5 Evaluation table of compost maturity

Issue	Evaluation (point)
Color (10)	Yellow ~ yellowish-brown (2), brown (5), dark-brown ~ black (10)
Texture (10)	Maintain original shapes (2), shapes are partially disintegrated (5), does not maintain original shapes (10)
Odor (10)	Strong excreta odor (2), weak excreta odor (5), compost odor (10)
Moisture (10)	Drips out between fingers when squeezed, 70% or more (2); heavily sticks to palm when squeezed, around 60% (5); does not stick much even when squeezed, around 50% (10)
Max. Temp. in Incubation (20)	50 °C or below (2), 50~60 °C (10), 60~70 °C (15), 70 °C or above (20)
Incubation Period (20)	Manure only: within 20 days (2), 20 days~2 months (10), 2 mo. or longer (20)
	Crop residue mixed: within 20 days (2), 20 days ~ 3 mo. (10), 3 mo. or longer (20)
	Woody materials mixed: within 20 days (2), 20 days~6 mo. (10), 6 mo. or longer (20)
Turnover (10)	Twice or less (2), 3~6 times (5), 7 times or more (10)
Forced Ventilation (10)	Absent (0), present (10)

(∗) If the total score is 30 or below, immature; 31~80, semi-mature; 81 or above, fully mature.
"mo." means month
Source: NARO, Harada (2003)

for these new materials, various additional treatment methods such as gasification, thermal decomposition, or carbonization are introduced to complement the conventional composting process. Most of these materials are intended for use in the farmland as nutrient additives or soil conditioners although the use may not be limited to agricultural applications.

The Fertilizers Regulation Act of Japan categorizes fermented compost or dried compost processed from sewage or night soil sludge as an ordinary fertilizer and defines the permissible levels of or the limitations on hazardous constituents based on plant hazard risk assessments. The Act also obligates the manufacturer to provide a list of the primary components contained in the composting products. In general, there are several standards for quality assessment if a specific recycled organic material is fit for use as a nutrient provider or soil conditioner for crop cultivation. For example, the content of fertilizer components such as nitrogen, phosphorus, and potassium or characteristic values such as the C/N ratio and pH can be used for evaluating the composting products. The decomposition rate of the organic components depends on the application conditions of the soils and the climate, and hence, predicting the effectiveness of these parameters precisely is difficult. In addition, the degree of nutrient intakes differs for each target crop, and thus the quantity of the fertilizer components remained in the soil will be different as well. Because on-site testing of the recycled materials in the field is time-consuming, a significant investment of time and cost is required to obtain the field test results. Therefore, it is desirable to establish a simpler and more convenient test method to assess the characteristics of recycled organic materials.

In the following sections, several evaluation methods that have been investigated in our research projects are introduced, and the criteria to assess the process of recycled materials from urban organic waste are evaluated, and how the information is dissimilated to manufacturers and consumers is discussed. Notably, the quality of nutrients in the recycled products used in the agricultural field is the most pressing concern for consumers, but there is also a strong demand for a comprehensive understanding of not only its component data but also its effect on the environment of the cultivation field and the crop growth.

2.2.3 Assessment of Recycled Materials Through Growing Tests

2.2.3.1 Test Methods

A series of experiments on establishing methods to evaluate the use of either composted or pyrolyzed human waste sludge as a new recycled material was conducted. Three types of compost materials used in the experiments included composted cow excrement, composted sludge, and pyrolyzed sludge; chemical fertilizer was used as a reference in the experiment. The composted cow excrement is fermented cow dung, whereas the composted sludge is fermented surplus sludge from night soil treatment plants, and pyrolyzed sludge comes from the same surplus night soil sludge except that it has undergone a 200 °C thermal decomposition procedure. Table 2.6 summarizes the components and characteristics of the tested materials (Ochiai 2005).

Brassica rapa var. *perviridis Brassica* (komatsuna) was cultivated as the test crop in these experiments. At each test section, a fertilizer composed of $N:P_2O_5:K_2O = 15:7:12$ (g/m^2) was applied as a basic fertilizer. Because organic matter decomposes rather slowly, very little of it will be decomposed into effective inorganic nutrients during the limited initial time of the test period. Therefore, the target materials were applied with the amount one to four times that of the regular

Table 2.6 Nutrient content in test material and soil

Test material	N (%DM)	P_2O_5 (%DM)	K_2O (%DM)	Mg (%DM)	Ca (%DM)	C/N (−)
Chemical fertilizer	7.06	5.39	1.99	0.42	4.61	0.08
Composted cow excrement	2.77	1.18	4.57	1.27	0.00	4.73
Composted sludge	4.38	10.82	1.59	2.14	0.02	5.49
Pyrolyzed sludge	4.70	7.23	1.76	2.07	0.02	6.12
Red soil	0.13	0.01	0.22	0.30	0.02	13.61
Black soil	0.47	0.03	0.14	0.20	0.06	13.58

fertilizer use. Also, at several sections a random mixture of two materials was applied with the application amount equivalent to one to four times the amount of normal nitrogen application. Plastic pots with 15 cm in diameter and 3 liters in volume were filled with the cultivation soil, and five crops at each test section were cultivated in isolated chambers in a netted room at the university research field (Fuchu, Tokyo).

2.2.3.2 Analysis Methods

Fertilizers and composts applied as cultivation nutrients are required to work effectively on the target crop; they need to remain in the soil for supplying nutrients to subsequent crops. In addition, the amount of organic matter being decomposed and absorbed into the crops is a significant criterion. In these experiments, the nutrient components contained in the harvested crops and left in soil were analyzed by using the following procedures. The three most fully grown crops were selected out of the five test crops at each section, whereas three soil samples were collected from the soil near the center of the cultivation pots with any roots removed. Nitric acid contained in the crop was determined by using the colorimetric method known as the Cataldo method. The phosphorus concentration was measured by using the colorimetric determination with vanadomolybdate acid, and concentrations of potassium, magnesium, and calcium were assessed using an atomic absorption spectrophotometer on samples processed with the wet ash method. The nitrogen concentration in the soil was measured by using the dry combustion method, and the other components contained in soil were analyzed using the same methods as for the crop analyses.

2.2.3.3 Test Results

Samples collected from various test sections are denoted by letters as follows: "a" for chemical fertilizer, "b" for composted cow excrement, "c" for composted sludge, "d" for pyrolyzed sludge, and "e" for no fertilizer applied. Similarly, the mixed fertilizer sections were marked with the combined letters of designation for components of the mixed samples, e.g., "ab" for chemical fertilizer with composted cow excrement. The level of fertilizer or compost applications was indicated by numeral "1"–"4" for every sample.

Figure 2.2 illustrates the crop yield at each section as well as the nitrate content measured in the leaf. In the single sample application sections, the increase of growth mass was observed to depend on the level of sample application; plants with chemical fertilizer application exhibited the most evidenced increase, whereas the crops grown with composted cow excrement showed a gradual weight increase. In the mixed sample sections, the combination of chemical fertilizer with composted sludge resulted in the most significant growth increase, followed by the combination of chemical fertilizer with pyrolyzed sludge. Although the combination of

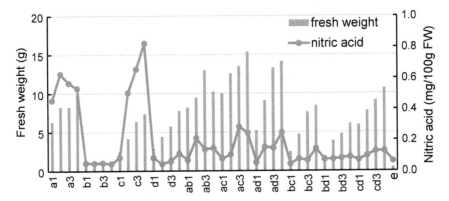

Fig. 2.2 Fresh weight and nitric acid content of komatsuna at harvest time

chemical fertilizer with composted cow excrement exhibited a faster growth rate than the sample with composted cow excrement alone, the mass growth was not necessarily proportional to the level of sample application. Takebe et al. (1995) conducted experiments with the same subject crop to investigate the effects of fertilizers on the growth and crop quality by using compost and sewage sludge as base fertilizers; the application level was three times the nitrogen amount normally applied. The results show that both the amount of nitrogen intake and mass growth at the compost application sections were half the results when standard chemical fertilizer was applied. The sections with sewage sludge application showed worse outcomes than the chemical fertilizer application sections.

High concentrations of nitric acid in crops can act as carcinogens. The nitrate ion concentration in the tested leaves was high for sections using chemical fertilizer and composted sludge. Particularly at the composted sludge section, the nitrate ion concentration increased in proportion to the quantity of sludge application.

Figure 2.3 shows the content ratios of other components: phosphorus, potassium, magnesium, and calcium. Phosphorus and potassium tend to increase according to the sample application amount at most sections except for potassium at the section with application of composted cow excrement, which exhibited a significantly lower potassium content. For magnesium and calcium, on the other hand, increasing application amount results in decreasing content ratios in the subject lamina except for the section with application of composted cow excrement. In addition, the relationship between the sample application amount and the content ratio became rather complicated at the sections with mixed application of composted cow excrement and other nutrient samples.

The nutrients remaining in the soil after a specific crop harvest is a significant factor to affect the next crop cultivated in the same field. Microorganisms living in the soil decomposed the organic matters left in the soil to release some nutrients contained in organic matter so that the released nutrients would become available to plant growth. Figure 2.4 shows the content ratio of the remaining nitrogen, phosphorus, potassium, magnesium, and calcium in the test section soil after the

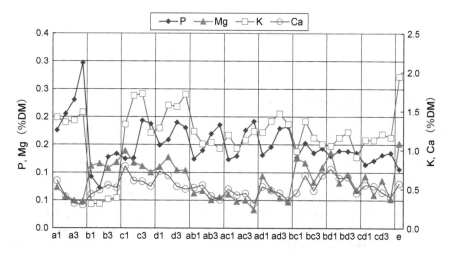

Fig. 2.3 Nitrogen, phosphorous, calcium, potassium, and magnesium in komatsuna at harvest time

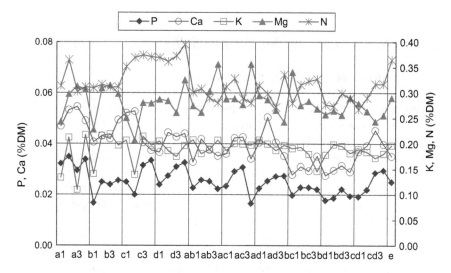

Fig. 2.4 Nitrogen, phosphorous, calcium, potassium, and magnesium in pot soil at harvest time

experiment period. Nitrogen showed an increasing tendency after the test period for the sections with separate compost sludge and pyrolyzed compost applications. At the mixed application sections, however, none of the sections exhibited the increase of nutrient remaining in the soil after cultivation, even for those with mixed application including either composted sludge or pyrolyzed sludge. Nishio and Oka (2003) conducted a plant growing test to investigate the dynamics of nitrogen contained in the soil applied with composted cow excrement and composted pig excrement mixed with chemical fertilizer. In the composted pig excrement plot, the nitrogen amount in the soil was greater after the experiment than the amount applied during the experiment, whereas the results with the composted cow excrement showed complete opposite results. The results of the present study have successfully confirmed the observation that the difference in dynamics of the remaining nitrogen in the soil is caused by the organic matter applied to the soil.

Most of the nutrient components other than nitrogen decreased after the experiment. Phosphorus and magnesium exhibited a high remaining content according to the quantity of compost application, whereas the post-experiment concentration of potassium and calcium did not show a significant correlation with the quantity of compost application. If the post-experiment amount of nutrient component remaining in the soil and the amount absorbed by the crop are deducted from the preexperiment amount of nutrient component in the soil, the amounts of nutrient components flashed out of the pots during the experiment were obtained. This quantity is referred to as "circumstantial flooding," and it was the highest in phosphorus at the composted sludge application sections, followed by the pyrolyzed sludge sections. Circumstantial flooding tended to increase at every section according to the sample application amount.

2.2.4 Notation of Growing Tests by Self-Organizing Maps

2.2.4.1 Structure of Self-Organizing Maps

As mentioned previously, organizing the data collected from growing tests into tables and graphs for detailed discussions and application analyses requires sophisticated mathematical technical knowledge. Regular users such as community farmers or ordinary citizens do not necessarily have extensive knowledge in these fields, and thus they need a simplified notation method to obtain useful application information. Here, author and colleagues introduce the "self-organizing map" (SOM) method, which can be used to present relative information in an organized fashion (Tojo et al. 2007).

SOM, a visualization technique for multi-dimensional information, was established by Kohonen (2001) and subsequently improved by many other researchers. SOM consists of two layers: the first input layer and the second output layer, with components called "units" that are aligned two-dimensionally (Fig. 2.5). Kohonen introduced the artificial neural network depicted by Eqs. (2.1 and 2.2) below. The units of the output layer, which express experimental sections, were organized according to its relationship distance by learning multiple data characteristics of input layer.

$$m_i(t+1) = m_i(t) + h_{ci}(t)\left[x(t) - m_i(t)\right] \tag{2.1}$$

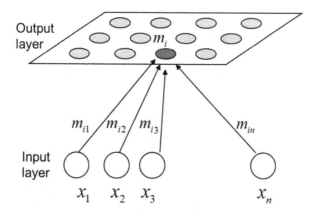

Fig. 2.5 Principle model of self-organizing maps

$$h_{ci}(t) = \alpha_0 \left(1 - \frac{t}{T}\right)$$ (2.2)

where $m_i(t)$: information vector at time t; h: proximity function; and x: input vector.

In this study, seven-dimensional data were used as the input layer: crop yield, concentrations of crop nitrate, phosphorus, potassium, magnesium, calcium, and residual soil phosphorus. Units were placed in a 50×41 grid on the output layer and were learned using the Gauss function as the proximity function.

2.2.4.2   Data Notation in SOM

Figure 2.6 shows the cultivation results for the target crop. The letters and numbers in the maps represent the sample fertilizers and application amounts at the test sections, whereas the distance between units that represents the test sections visually shows the general data similarity. The unit color represents the component value: red for "high," green for "intermediate," and blue for "low." This map shows that the composted sludge resulted in a similar data leading to the chemical fertilizer; particularly sections with three-time and four-time applications of composted sludge had similar results as the section with a one-time application of chemical fertilizer. Magnesium is known as a guest element for plant growth, and the map clearly shows that magnesium concentration tends to be high in the sections with low mass growth. As the effects of the mixed use of recycled materials is concerned, the map shows a general evolution that the sections with mixed use of chemical fertilizer and compost show data close to the sections with application of only chemical fertilizer; yet the combination of composted sludge and pyrolyzed sludge is also relatively close.

SOM can visually present the proximity of sample characteristics thus enabling an easy grasp of characteristics of the recycled materials based on its proximity to

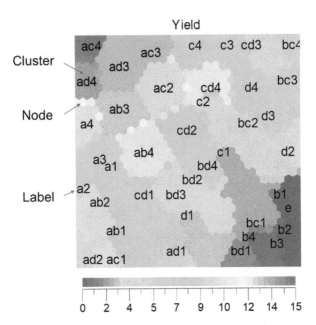

Fig. 2.6 Self-organizing map (SOM) illustrating yield of komatsuna cultivation test. Cluster shows a group of nodes (units) that have similar characteristics. Label shows experimental section of test material and amount

chemical fertilizer and composted materials of general use in the agricultural field as well. This type of data presentation tool should be more commonly used in the future as a communication tool between manufacturers and consumers or among consumers.

2.2.5 Communication on Recycled Material

Organic waste discarded in cities should be processed according to its appropriate forms into highly recyclable products in order to establish a nearly complete recycling society. This is a challenging task that requires all of our knowledge and wisdom to envision the type of recycling system to be included in the fundamental improvement of urban infrastructures. One of the most crucial targets in the use of recycled products is in agricultural applications, specifically, in farming practice. Knowledge and information concerning the impact of recycled products on the processes of decomposition, mineralization, and crop intake, as well as their remaining residues in soil, will be the most fundamental concerns for promoting the practice of environmentally friendly agriculture or organic farming.

   Both the manufacturers and consumers of recycled products must share the same knowledge and understanding of the products being produced or used. Otherwise, a few specific nutrient components may concentrate in the soil and eventually leak into surface runoffs or groundwater causing significant adverse environmental impacts. A clear and easy-to-understand standardization and notation method to characterize the products is required for both manufacturers and consumers.

References

Kohonen T (2001) Self-organizing maps. Springer-Verlag, Berlin Heidelberg

Kurita E, Yamamoto T, Shigeta T (2010) The potential of agricultural land management by the user organization of the allotment gardens in peri-urban area: a case of allotment gardens for kitchen garbage recycling, Kitamoto city. J Rural Plan 29(3):349–352

Harada Y (2003) Composting of organic waste and its quality. NARO website. https://www.naro. affrc.go.jp/training/files/20031006-08-1.pdf. Accessed 25 Aug 2018

METI Agency for Natural Resources and Energy (2015) Feed-in tariff of renewable energy of Japan. http://www.enecho.meti.go.jp/category/saving_and_new/saiene/kaitori/kakaku.html. Accessed 25 Aug 2018

Motghare KA, Rathod AP, Wasewar KL, Labhsetwar NK (2016) Comparative study of different waste biomass for energy application. Waste Manag 47:40–45

MAFF (2018) Discharge of food waste and executing rate of recycling. http://www.maff.go.jp/j/ shokusan/recycle/syokuhin/kouhyou.html. Accessed 25 Aug 2018

NARO (2010) List of basic data for grasping the biomass amount present in a region. http://www. naro.affrc.go.jp/nkk/introduction/files/basic_data.pdf. Accessed 25 Aug 2018

Nishio T, Oka N (2003) Effect of organic matter application on the fate of 15N-labeled ammonium fertilizer in an upland soil. Soil Sci Plant Nutr 49(3):397–403

Ochiai S (2005) Assessment of recycled organic waste materials on crop growth and soil. BS thesis, Tokyo University of Agriculture and Technology (in Japanese)

Sano H (2007) Biomass combustion. J High Temp Soc 33(1):3–8

Tojo S, Ochiai S, Tanaka H, Suzuki S, Watanabe K (2007) An evaluation method of various recycled organic materials using a self-organizing map. Jpn J Farm Work Res 42(4):189–198

Takebe M, Ishihara T, Matsuno K, Fujimoto J, Yoneyama T (1995) Effect of nitrogen application on the contents of sugars, ascorbic acid, nitrate and oxalic acid in spinach (Spinacia oleracca L.) and Komatsuna (Brassica campestris L.). Japanese Journal of Soil Science and Plant Nutriotion 66: 238-246. (in Japanese)

Yamagishi Y, Kurihara M (2012) An estimation for CO2 reduction amounts by various recycling ways of pruning waste. Pap Environ Inf Sci 26:237–242

Chapter 3
Carbonization of Wasted Biomass and Carbon Sequestration

Pisit Maneechot, Prapita Thanarak, and Haytham M. El Sharkawi

Abstract Charcoal primarily consists of carbon, which is the remains of hydrocarbons after its moisture and other volatile matters are driven out of the material by using pyrolysis. The pyrolysis process is carried out in which the biomass is heated to a temperature high enough in the absence or with limited amount of oxygen for eliminating the moisture and the volatile matter. Biochar is made by pyrolysis at low temperature around 400 °C and intended to be used as a conditioning agent to remedy a rough soil. A broad aspect of the impact includes physical, chemical, and biological properties of the soil as well as the yield of produce from the conditioned soil. Other benefits include enhanced carbon content and storage for carbon sequestration, boosted plant nutrients, amended soil structure, and improved hydrology management. Several variations of smokeless kilns have been produced. Using electric heating results in near smokeless carbonization but makes the product more expensive in the end. Recapturing and treating the flue gases require intricate design and construction that makes the kiln more expensive to construct.

In Egypt, the new mechanized metal charcoal kiln is expected to improve local air quality by reducing emission of greenhouse gases, especially carbon monoxide and carbon dioxide, and protect the local groundwater quality and land resources as well as vegetation and agricultural production from tar emissions. The sandy soil is incapable of retaining irrigation water and nutrients because it contains less amount of organic matter. Farmers apply organic matters or composts to improve the soil fertility. New lands in the Delta of Egypt are characterized by the absence of many diseases and contaminants that may be suffered by the old farmland in the same

P. Maneechot (✉)
Excellent Center for Asia Pacific Smart Grid Technology (APST), Naresuan University, Phitsanulok, Thailand
e-mail: pisitm@nu.ac.th

P. Thanarak
School of Renewable Energy and Smart Grid Technology (SGtech), Naresuan University, Phitsanulok, Thailand
e-mail: prapitat@nu.ac.th

H. M. El Sharkawi
Agricultural Research Center (ARC), Giza, Egypt

© Springer Nature Singapore Pte Ltd. 2020
S. Tojo (ed.), *Recycle Based Organic Agriculture in a City*,
https://doi.org/10.1007/978-981-32-9872-9_3

region. Using biochar and other value-added products instead of compost is one of the promising ways because it does not include the plant root diseases such as nematodes and fusarium and prevents from excessive application of pesticides and the resulting high production costs.

3.1 Carbonization of Biomass

3.1.1 Charcoal, Biochar, and Activated Carbon

Charcoal, biochar, and activated carbon are composed of basically the same materials; the distinction among them depends largely on their applications and the production conditions. Charcoal primarily consists of carbon, which is the remains of hydrocarbons after its moisture and other volatile matters are driven out of the material by using pyrolysis. It is sometimes described as the first synthetic material or energy source produced by mankind (Antal and Grønli 2003). Other components such as ash and some residual volatile matters exist in charcoal (Goulart et al. 2017; Maroušek et al. 2017) (Antal and Grønli 2003), which is mainly black in color. Charcoal has been used as a fuel source or simply a source of heat energy since the early generations of mankind (Deforce et al. 2013; Raab et al. 2015; Goulart et al. 2017) and still continues to be one of the dominant energy sources in both developed and developing nations around the world (Mulenga et al. 2017). Charcoal or biochar may be applied in many facets of human activities, but they are used primarily for domestic and industrial applications (Aabeyir et al. 2016; Raab et al. 2015).

The production of charcoal is carried out by using the pyrolysis process in which the biomass (plant and animal matter) is heated to a temperature high enough in the absence or with limited amount of oxygen for eliminating the moisture and the volatile matter (Pereira et al. 2017). This results in an incomplete burning to convert the biomass into charcoal that has higher heating value than the original material. The type of raw material used basically gives rise to the type of charcoal produced. While charcoal may be produced from the raw biomass directly, it can also be produced from preprocessed or refined biomass.

Biochar and charcoal are synonymous with a slight difference in their applications. Charcoal is primarily used as a heat energy source, and biochar has been mainly applied to remove polluting materials from air, water, and/or soil (Lehmann and Joseph 2015). The conditions to produce charcoal and biochar remain virtually similar, however, biochar is pyrolyzed at relatively higher temperature by using modern pyrolysis methods. Biochar has been applied in environmental remediation as adsorption/desorption and oxidation/reduction techniques to achieve carbon sequestration, nutrient exchange, and water treatment, among others. These biochar-based techniques have been found to result in many benefits such as increasing nutrient availability, microbial activity, water retaining capacity, and soil organic

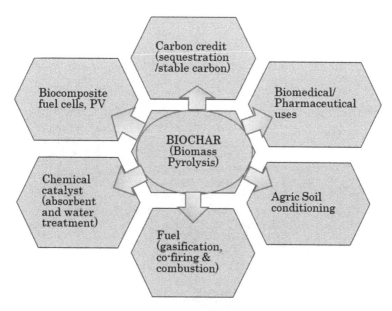

Fig. 3.1 Schematic diagram of biochar utilization and application

matter availability, as well as adsorbing contaminants in soil and thus conditioning the soil to enhance crop production. (Fig. 3.1). Although these soil and water remediation and reconditioning techniques using charcoal have been studied recently, some ancient agronomical practices already applied the principle of these techniques a long time ago by using the slash-and-burn method in farming (Gay-des-Combes et al. 2017) till nowadays. This practice is found to add biochar to the soil as carbon sequestration and soil conditioning that farmers have attested the enhancement of crop yield (Lehmann and Joseph 2015).

Biochar, activated carbon, and charcoal are all pyrogenically synthesized materials (Hagemann et al. 2018) so that a clear distinction among them is difficult to define although charcoal can be distinctly classified. The benefits of using them as materials to remedy environmental issues have been studied in addition to using them as an alternative energy source (Hagemann et al. 2018; Richards 2016; Wu et al. 2017). Unlike charcoal, activated carbon (AC) can be made from a variety of materials including wood, peat, lignin, bituminous coal, lignite, and petroleum residues (Fig. 3.2). Production of AC involves a two-phase process: the first carbonization and the follow-up continuous heating and the second phase. During the first carbonization phase, amorphous materials are burnt off by dehydrating the material at approximately 170 °C. After a continuous heating, the evolution of CO, CO_2, and acetic acid begins and completes at a temperature between 400 and 600°C. The product at this point is an intermediate product. During the second phase, the intermediate product is decomposed at a temperature between 750 and 950 °C that causes pores to widen, thus creating macropores to enhance the carbon adsorptive powers (Hagemann et al. 2018) of the AC.

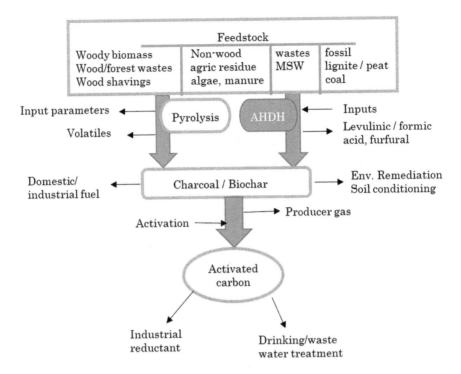

Fig. 3.2 Pyrogenic carbon production and applications

3.1.2 Carbonizing Techniques and Applications

Woody biomass consists mainly of water, cellulose, and lignin, as well as other minerals, acids, and volatile materials. The tightly bound lignin and cellulose form the basic structure of the woody material to hold molecular water. A freshly cut wood, regardless of its type, may contain about 40–100% of the total mass as compared with the oven dry weight. After being seasoned to the best practical standard, the wood still contains some entrained and absorbed moisture that accounts for about 12 to 18% of the total dry mass. Carbonization is a destructive distillation process to convert hydrocarbon organic matters primarily into carbon (Byrne and Nagle 1997) with some traces of other elements. Carbonization is considered a pyrolytic process that involves several reactions to achieve the end result containing largely carbon. Thus, the carbonization of wood into charcoal is to convert the hydrocarbon material into carbon, which is to be used as fuel and adsorptive materials. Carbonization of wood into charcoal can also be described as a spontaneous breakdown of the wood structure to form charcoal as the residual material.

3.1.2.1 Direct Combustion

Most traditional kilns and charcoal burners used the direct combustion method to carbonize wood or biomass. With this method, some of the raw materials is burned off to provide the necessary heat for completing the pyrolytic process. In such instance, the yield is low because air/oxygen is initially supplied to start the combustion process until the evolved gases begin to take over the burning process; the air supply is then cut off or restrained. As indicated in Fig. 3.3, the wood, which is also known as "charge" in the reactor, is ignited with air supply being continued until the produced gas is enough to sustain the pyrolytic process. Thus, portion of the charge is consumed and subsequently turned into ash. The direct combustion process tends to slow down the combustion resulting in longer recovery time thus adversely affecting the quality and quantity of output. The advantage of this method is that no additional fuel is needed for the process; further, the construction and operation of the process are relatively simple. Hence, ground pit and masonry kilns are suitable for rural biochar production because the capital cost is low when the product quality and process time are not of prime concern.

3.1.2.2 Indirect Combustion

The indirect combustion method is used to enhance the yield quality and quantity of the product in addition to saving process time. In this method, external heat source is used to pyrolyze the biomass with limited oxygen supply throughout the process; the fuel for heating in the combustion operation is separated from the charge. Most

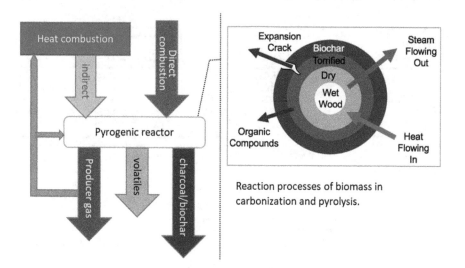

Reaction processes of biomass in carbonization and pyrolysis.

Fig. 3.3 Schematic diagram of carbonization

retorts and modern kilns use the indirect method to carbonize biomass into charcoal. The design uses a separate chamber to provide the heat to the charge loaded in another chamber. The heat from the source and the gases are transferred to the charge across a grit of channel by heat conduction and convection. The escaping gas from the charge is redirected to the heat source to sustain the combustion and raise the temperature of the charge as well. This method allows the highest percentage of yield because almost all inputs are directly converted into the output carbon. The nearly constant temperature and ensured air supply warrant the output quality. The downside of this process is that the initial capital investment of kiln can be high and the process needs to be strictly controlled.

(i) Carbonization Practices

The main purpose of carbonizing biomass is to separate components of biomass into carbon known as charcoal, gas, and condensable volatiles. The resulting charcoal has higher energy density than the original raw biomass because the "unwanted" components have been removed to increase the heat value. To achieve an optimal yield, the process must be carefully controlled despite the variation in methodology and practices reported in literature (Fig. 3.4). The principle is common, but the practices are slightly modified to satisfy the objectives such as quality of charcoal, reduced emission of gaseous by-products, and duration of reaction time, among others. The general procedures are discussed in the following sections.

1. Dehydration

Fresh wood contains sap and the seasoned wood also would contain some amount of absorbed moisture. The absorbed or entrained moisture must be eliminated or reduced to a possible minimum so that the subsequent processes can occur (Antal

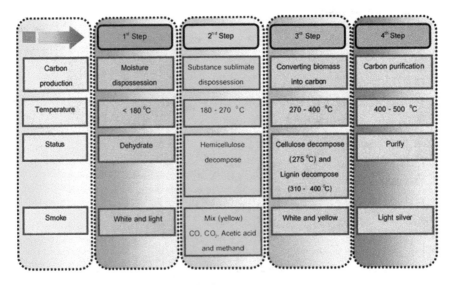

Fig. 3.4 Carbonization and charcoal production process

and Grønli 2003). The endothermic process of using heat to reduce this moisture in the wood is known as "dehydration," which is also referred to as the "desorption of adsorbed water" or "moisture dispossession" process. It occurs when the wood temperature is raised from ambient temperature to 100–110 °C, and this process can continue when the temperature is raised to and maintained at 180 °C (Tang and Bacon 1964); the moisture thus escapes in the form of vapor. Water vapor tends to limit the rate and efficiency of combustion so that its elimination from the biomass enhances the carbonization process. Thus, the biomass must be pretreated by drying with any suitable and economical method to enhance the rate of carbonization (Xie et al. 2009). A well-seasoned wood is easy to carbonize to achieve the desirable results in a reasonable time. Because the economic benefits of charcoal production are relatively low, the wood seasoning is carried out mostly by using air and sun. When the wood drying is done in a controlled environment, the wood can be dehydrated to an appreciable level so that the carbonization process can proceed with relatively low energy input requirement.

2. Substance Dispossession and Emission

The endothermic reaction continues from the dehydration stage to the wood decomposition stage at temperatures up to 170 °C. When the cellulose structure is split up at 240 °C, components such as tars, organic acids, CO, and CO_2 as well as chemically bonded water are released from the wood cells. These released gases are combustible; they provide the fuel to raise the wood temperature to between 230 °C and 250 °C that transforms the wood to a reddish-brown nonbiodegradable material.

3. Biomass Conversion and Purification

"Torrefaction," which is a chain scission or depolymerization process to break up the C–O and C–C bonds, occurs at temperatures between 240 °C and 400 °C. The end product is black in color with little strength. The later stages of this process overlap with the carbonization process that starts at temperatures higher than 350 °C and completes with temperatures reaching the range of 450–550 °C (Xie et al. 2009). This stage is also referred to as "aromatization" in which a graphite layer is formed. Carbonization occurring at a lower temperature has been found to result in better yield, but the end product has low grade because it contains high percentages of acidic, tars, and volatiles. Although some researchers suggest that a slower rate of pyrolysis yields charcoal of better quality, others suggest that there is no noticeable difference between products whether they are produced under normal conditions or not (Antal et al. 1990, Chan et al. 1988). Additionally, the breakdown temperatures are 200–260 °C for hemicellulose, 240–350 °C for cellulose, and 280–500 °C for lignin (Byrne and Nagle 1997).

(ii) Carbonization Parameters

Among the parameters that are found to correlate closely with the yield of pyrolysis reactions, the nature of the input wood feedstock is of great importance. Properties such as density, moisture content, chemical composition anatomy, and dimensions will affect the quality of the charcoal produced as well. Charcoal density is largely determined by the wood feedstock used as the raw material. Wood with higher density will produce a denser charcoal that is more resistant to parallel compressive force and less friable (Chrzazvez et al. 2014). Even though the general charcoal produced at increasing temperatures may compromise in density, denser wood feedstock still results in more condensed charcoal. Additionally, the moisture contents of wood with different ages may vary; wood with higher moisture content requires more heat energy to complete the carbonization process, thus resulting in cracking both internally and externally in the charcoal produced (Pattnaik et al. 2018, Rousset et al. 2011) due to a high degree of shrinkage caused by excess steam pressure. The major components of wood such as cellulose, hemicellulose, and lignin undergo mass loss during carbonization. While cellulose is found to lose almost 60% of its mass, hemicellulose and lignin lose greater percentage of mass; thus the percentage composition of these components in the wood will significantly determine the yield of the charcoal (Pereira et al. 2013). Among the many factors that affect the carbonization process, temperature, pressure, and heating or reaction rate are the three most predominant factors.

1. Temperature

Because pyrolysis is a thermal process, temperature is a determining factor on the process efficiency. Figure 3.4 outlines the impact of temperature on carbon content, yield, hydrogen content, electrical resistivity, specific gravity, and water sorption ability. Charcoal is seen to form around 300 °C to reach a total yield of approximately 50%. The rate of charcoal yield drops rapidly at higher temperature until 600 °C. Beyond 600 °C, the declining rate slows down to a constant final yield. The carbon content consistently increases to more than 96% at 1000 °C, whereas the hydrogen content reduces from less than 10% to almost 0% beyond 600 °C, and the specific gravity increases sharply and stabilizes at around 700 °C and beyond. The electrical resistance decreases to less than 10 Ω/cm at 800 °C, and water sorption capacity decreases to about 50% beyond 1400 °C (Dorner-Reisel et al. 2017).

Higher temperatures are associated with increasing compressive strength when the temperature is maintained between 300 °C and 900 °C as a result of the increasing quantity of fiber per unit area. Increasing porosity and pore size and changing porosity distributions and pore shape as a result of high carbonization temperatures to extract more volatile matter from the mass have been reported in literature (Tintner et al. 2018). Castro et al. (2018) also confirm that increases in charcoal density, resistance, and surface rough are associated with higher temperatures of carbonization.

2. Heating Rate

The rate of applying heat energy to the charge in the reactor has been found to influence the yield of charcoal. Pyrolysis with fast or intermediate rates, which occurs within a few seconds, mostly results in producing condensable liquids with minimal solid char content (Maruyama et al. 2018). Slow pyrolyses on the other hand, occurring in 10–60 min or longer time, cause carbonization, combustion, torrefaction, and gasification of the raw material. The fast heating rate results in charcoal with increasing friability and decreasing compressive strength (Chiang and Pan 2017). The slow heating rate has been found to cause smaller cracks because of smooth and mild drying process as well as less porosity (Villota et al. 2017). Fast rate is also responsible for dimensional alteration of the charcoal as a result of rapid shrinkage caused by escaping volatiles.

3. **Pressure**

Increasing pressure in the reactor vessel during carbonization results in the condensation of volatiles in the solid matrix. This leads to the formation of secondary charcoal with more stable structure and higher yielding of carbon. Elevated reactor pressure enhances easy and uniform heat transfer throughout the vessel thus ensuring uniform combustion and quality output. The quality of the charcoal produced in the reactor with elevated pressure and temperatures is good enough for industrial use such as in steelmaking.

(iii) **Charcoal Application for Energy**

Charcoal provides heat energy for both domestic and industrial applications. It has been the major source of heat for all countries and cultures worldwide (Lennox 2017) and remains the most preferable choice of heat source in almost all developing countries nowadays (Monteiro et al. 2017). Charcoal was also used in the US, Brazilian, and European iron-smelting industries before the advent and adoption of coke and coal. Charcoal was known to spearhead the fuel industry in the eighteenth and nineteenth centuries; it has been used for household cooking throughout the world because of cultural and lifestyle reasons or its affordability over other types of fuel. In some cultures, some traditional foods need to be processed by using charcoal to enhance the flavor or taste such as the use of briquette and charcoal for grilling barbeque and cooking, among others.

(iv) **Charcoal Application in Agriculture**

Biochar, which is a remnant of the pyrolytic process, is intended to be used as a conditioning agent to remedy a contaminated soil. The impact of biochar on agricultural application is illustrated in Fig. 3.5. A broad aspect of the impact includes physical, chemical, and biological properties of the soil as well as the yield of produce from the conditioned soil. The alteration of the soil biological and chemical properties improves the soil quality (Dokoohaki et al. 2018) and enhances plant growth and yield (Sara and Shah 2018). Other benefits include enhanced carbon content and storage for carbon sequestration, boosted plant nutrients, amended soil structure, and improved hydrology management (Dokoohaki et al. 2017). Additionally, biochar modifies the soil basic composition to decrease soil acidity and also cause nutrients and light organic molecules contained in the soil to be more amenable to plant growth.

Other advantages of applying biochar to the soil include enhancing aeration, increasing macropores to accommodate more microorganisms, and reducing soil bulk density (Dokoohaki et al. 2017). Before the advent of the intentional beneficial

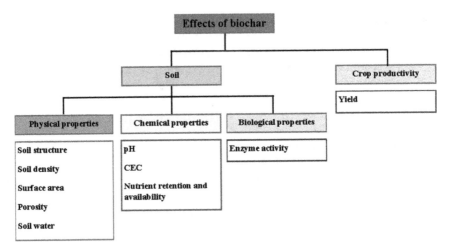

Fig. 3.5 Biochar application in agriculture

application of biochar to the soil, ancient traditional farmers noticed the impact of practicing slash-and-burn method to prepare land for farming. Research results suggest that certain minerals such as phosphorous are returned to the soil although the overall adverse impact on the environment of this method is huge (Gay-des-Combes et al. 2017). Biochar is also found to increase soil-specific surface area because of its own relatively larger surface area, thus enhancing the retention of nutrient and water in the soil available for plants to access. As biochar is porous in nature, its application enhances the porosity of the soil for better aeration to improve oxygen supply to both plant roots and microorganisms (Dokoohaki et al. 2018, Wu et al. 2017). In several agricultural field trials, biochar has been found to improve crop growth significantly. Haruthaithanasan (2003) reported increases of plant fresh weight and height when biochar was applied to soil at a ratio of 100:0–25:75 in a field trial of growing garden vegetables. Increases in leaves and branches were also observed in the same trials.

(v) **Wood Carbonization Value Chain**

The value chain of wood carbonization along energy, agriculture, and other non-energy applications is presented in Fig. 3.6. On the energy linkage, wood is carbonized to produce charcoal, condensable liquids like vinegar, wood gas, and biochar. Charcoal and wood gas can be used directly by both domestic and industrial sectors of the economy as fuels. Most rural and urban dwellers rely on charcoal for cooking and other heating applications; the wood gas can be used for small-scale electricity generation.

The agricultural value chain applies vinegar and biochar for remediating soil, water, and environment. Vinegar is used as insect repellant by farmers and as nutrient supplements to plants. Biochar is used as soil amendment to improve yield of agricultural produce and chemical reagents for water treatment (Lehmann

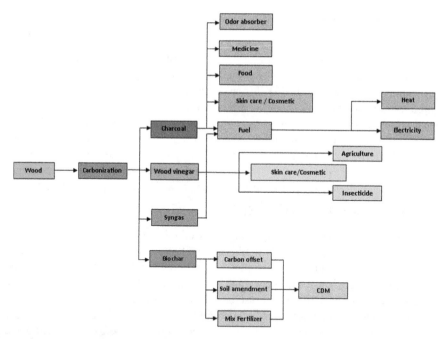

Fig. 3.6 Charcoal/biochar value chain

and Joseph 2015). With soil amendment resulting in carbon sequestration, a CDM (Clean Development Mechanism) application can be achieved with biochar (Platt et al. 2018). Other non-energy value applications include cosmetics, medicine, and food.

For a full potential of the value chain be realized, a well-designed supply chain and business models must be established. The business model should be fashioned along the energy, agricultural, and allied applications; it considers a robust and reliable supply chain generally made up of producers, transporters, and marketers or value addition. With a fully developed business model, the full potential of wood carbonization can be realized along the value chain. The downside of this optimism is that less developed countries are involved much more in the charcoal industry and the challenges they face are enormous. Technologies are low for the conversion of these energy sources into full benefits; thus, the associated business models are not always strong and robust (Habte and Hector 2017). On the supply chain, the low technology impedes a full conversion of wood/biomass into usable products, thus resulting in a wasteful and inefficient process. For instance, most gases and liquids are lost due to the use of low conversion technology (Puentes-Rodriguez et al. 2017), and transporting and marketing the finished products pose challenges as well. Therefore, low technology also impeded high add-on values. In addition to these factors that directly impede a full realization of the potential, other factors such as lacking appropriate policies and supports by institutions cause value loss in the chain.

3.1.3 Type of Charcoal Kiln in Asia

As intense fuel requirement for domestic and industry applications becomes prevalent, new and efficient methods were devised to produce charcoal in quantity with efficiency (Woolf et al. 2017). Kilns are such devices/structures used nowadays to produce quantitative and qualitative charcoal for commercial applications. The normal processes of burning is to subject wood or any organic materials to high temperatures with adequate amount of oxygen or air for degrading the wood or the material into volatiles and ash. Charcoal and biochar on the other hand are produced by carbonization or pyrolization to decompose the wood thermally into carbon matter primarily and some volatiles; the carbon can be used as fuel to provide heat energy.

(i) Ground Pit and Mound/Pile Kilns

Charcoal can be produced by using a pit or kiln to produce the product with expected quantity and quality. The skill of the pit or kiln operator as well as the raw materials used and pit or kiln operational conditions will determine the characteristics of the product. The basic operational conditions of sufficiently high temperature and limited supply of oxygen must be adhered to. Historically, the ground pit method was used for charcoal production with a pit dug out in the ground preferably near the source of raw material such as near or in a forest where colliers will make their temporal home to supervise the operations (Svedelius 1875). This method is found to be convenient and simple to operate although labor requirement is intensive (Raab et al. 2015).

The pit method of charcoal production uses one of the simplest kiln designs requiring minimal tools and logistics for the construction. This method is intended not to be permanent; it may be used for a few months to a maximum of 1 year. Simple tools such as machete, pickaxe, mattock, and a shovel are required to dig out the pit. The site is first cleared of debris, and the pit dimensions are marked on the ground. The pit is then dug out according to the specified depth along the outlines marked with the scooped earth piled along the edges to be used later in the subsequent operations. For a more permanent site, the sides of the pit are rammed to prevent cave-in if the earth is loose or the side walls are cut outward as in other cases (Fig. 3.7).

Fig. 3.7 Charcoal pit dug in the ground (artistic impression)

The wood or charge to be processed is placed in the pit; it is usually isolated and insulated from the wet soil with grass, straw, or base log wood. In addition to being used as fuel, the grass, straw, or base log wood also creates draft to provide passages for gas to circulate during the burning process. The wood is placed in layers around a middle chimney to allow draft. It is ignited from the top to burn downward throughout the bed; the wood may also be ignited from below (Smith et al. 2017). In typical pit burners, after charging, the top of the wood pile is covered with twigs, straw, and leaves or charcoal dust with a final layer of sand. These covering materials are integrated with some of the charge serving as the necessary fuel to provide heat while the sand layer limits the amount of oxygen accessible by the burning system. Additionally, the sand layer also helps exhaust the gas and vapor trapped inside the kiln to keep them from impeding the burning process.

1. Feature of Various Pit Kilns

(a) Pile/Earth Mound

In the pile/earth mound method, the charge is piled into a mound-like pile on a cleared ground surface without digging a pit. The entire pile is then covered with brush, leaves, and sand to keep heat inside and also prevent the entry of air. One obvious advantage is that costs and energy associated with digging the pit are saved. In addition, this system ensures that any amount of wood gathered at one time can be entirely processed in one batch because the pile or mound does not have any restrictions in dimensions. Further, discharge of the final product is also relatively easy because all materials are placed on the ground surface. One major disadvantage is that this method takes a significant effort to cover the charge completely and securely; constant supervision is necessary to continually fill in the cracks that may develop in the cover. Figure 3.8 shows the outside appearance and the cross-section of a covered charge ready to be burned. In this case, a chimney is installed on top of the pile to allow the gas to escape. Although this system is simple in construction, it has low efficiency and poor charcoal quality because heating is sometimes not uniform. In both pit and mound/pile systems, charcoal is

Fig. 3.8 (a) A pile burning and covered, (b) cross-section of a pile ready for burning

likely to be mixed with the dirt thus increasing the percentage of impurities contained in the charcoal. This system also suffers adverse impact from natural elements such as rain and strong winds.

The pile/earth mound approach is practiced in many parts of Asia, Africa, and other less developed regions of the world by small-scale colliers who are mostly peasant farmers and hunters. Charcoal production becomes a secondary engagement when farmers are less busy from their usual farming routine. Features of pit and other types of kilns are summarized in Table 3.1.

(b) **Mud Kiln**

The mud kiln is a type of the pile system developed to improve the efficiency by constructing a more permanent chamber with mud or clay for burning the charge into charcoal (Fig. 3.9). In this construction, a strong clay or mud mixed with materials like palm fronds is used to construct an oven-like structure in the form of a wattle-and-daub structure to house the charge during burning. With this system, wind and rain have less effect on the process so that a cleaner charcoal can be produced. The heat is evenly distributed throughout the burning chamber to achieve a more uniform burning in addition to controlling air supply and heat escape more efficiently. The disadvantage is that wood must be cut into exact sizes to fit the oven interior dimensions, and material charging and discharging become more difficult than in the pile system.

(c) **Masonry Charcoal Kilns**

A more permanent and robust kiln, known as the "brick and stone masonry kiln," is designed to last longer and to produce better products at higher efficiency. This variant of kiln comes in different designs, which are affected by geographical location and the experience of the collier, as shown by the examples in Fig. 3.10. Bricks are preferred material to construct the kiln because it is cheaper and can withstand high temperatures without significant damage in addi-

Table 3.1 Comparison of kiln types

Burning method	Advantage	Disadvantage
Pit type	Cheaper and easier to construct	Product comes dirty and dusty
	Give higher output volumes	Affected by weather
	Constructed anywhere/near wood source	Ignition consumes more wood and does not burn uniformly
		Less efficient system because of leaks in structure
Kiln type	Not affected by weather and can be operated anytime avoiding stockpiling	Expensive to construct
	Product is always fresh and clean	Output per batch is lower
	Located where it can easily be attended to	Wood source must be transported to burning site
	Easy and cheaper cost to operate	

Fig. 3.9 Mud kiln with an open "door"

(a)   **(b)**

Fig. 3.10 Brick masonry kiln, (**a**) conical brick, (**b**) Iwate kilns

tion to being a good insulator to keep heat within the kiln. The basic shape is either rectangular or round with a flat or a dome top cover; the kiln can also be constructed in a conical or a beehive shape depending on the designer's ideas and experience. Some basic features are common to all kiln shapes. The main feature of a kiln is a door, thick walls, and a chimney. The door is used for charging and discharging ports, and in some designs, it serves as the ignition port. The chimney serves to exhaust smoke out of the kiln; some designs also use the chimney to set fire for ignition. The kiln walls are made thick to insulate the kiln and also prevent entry of excess oxygen into the kiln that may compromise the carbonization process. For this reason, double-layer brick walls were preferred in some kiln designs but later found to be difficult to control the process and mend cracks that may develop. In addition to using braces to reinforce the kiln, round kilns are also constructed with thick kiln-girding flat iron bands to restrict the wall expansion during the carbonization operation.

(d) **Drum Kiln**

The 200-liter oil drum has been used lately to construct charcoal kilns for various private and commercial charcoal productions. The advantages of using the drum include portability, efficiency, easy recovery of volatiles and liquor, ease of operation, as well as clean charcoal production. Disadvantages include heat loss, uncomfortable working environment due to excess heat radiated from drum, danger of explosion if not well managed, and short lifetime of less than 1–2 years due to rust of the metal drum. A typical production time is 1 day with charcoal output between 15 and 20%. A number of variations of the oil drum kiln have been proposed for more efficient production of charcoal and biochar.

Figure 3.11 shows a typical oil drum kiln construction. The basic concept of the drum kiln remains almost the same for all variations of design. The components include a cover with a hole or more likely a chimney, as well as an airtight base with opening for ignition, draft, and collection of volatiles. Some variations attach a number of right-angled tubes to the base to serve as chimney and for collection of volatiles. The cover is mostly for charging and discharging. The drum is placed on the ground or supported on stone pillars during firing. Other design variations include some auxiliaries for particular functions.

2. **Kiln and Retort System**

Burning the wood to produce charcoal or biochar has been found to suffer low yield of products with low quality. A retort is proposed to improve efficiency and product quality by using the hot gas from burning fuel to char the wood for improving the percentage of carbon content in the product. Using this principle of operation, retorts can be applied to process materials including small branches and twigs, grass, wood shavings, and just any types of organic materials into charcoal with good quality.

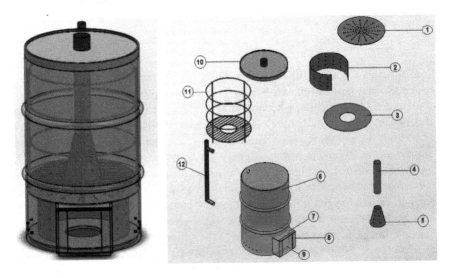

Fig. 3.11 Constructional drawing of the drum kiln

Because retorts use the escaping gases as fuel, the rate of pollution is kept minimal as compared with other types of kilns or traditional methods to make charcoal. Wood gases and condensable liquids can be easily recovered from retorts for other applications. A typical retort like the Hookway type can output charcoal containing 90–94% carbon with 50% efficiency. Retorts can be made of firebricks, concrete casting, and metal sheets or their combinations. With good conditions and resources, the retort can complete a charge within 8 h and then cool the product down in about the same time. Most metal retorts are made to be mobile so that they can be moved from place to place. An example retort is the one made by Dr. J.C. Adam of Technical University in Zvolen (Das et al. 2018).

Several inventions of the kilns have been demonstrated and evaluated at various locations of the world, and some have been adopted for private and commercial uses. Some notable places in Africa where there have been extensive research conducted and kiln designs proposed include Kenya, Senegal, Malawi, Mali, and Ghana (Smith et al. 2017). Metal, brick, and stone kilns have evolved and been redesigned in Africa to improve efficiency (Ngatia et al. 2016, Smith et al. 2017).

(ii) **Community-Type Kiln**

Until recently, most kilns were designed for commercial and industrial purposes. As indicated earlier, in Vermont, Pennsylvania, and New York as well as in Europe, the kilns were used to supply charcoal for smelter and refinery operations of various metals in the eighteenth and nineteenth centuries. In most parts of Africa and other developing worlds, the earth mound and pit-type kilns remain the predominant types for both commercial and business uses. Most commercial and industrial kilns have been constructed using bricks. These brick-type kilns, which have been largely operated by privately owned business or enterprise, are meant to yield profits for the kiln operators and all subsidiaries in the value chain. Community-type kilns have been mostly operated by researchers and academic institutions for conducting researches to improve kiln construction and operation and contributing to community development.

(iii) **Commercial-Type Kiln**

Most charcoal-producing enterprises are meant for business applications except a few that are for household consumptions. The business-type kilns take the shape of all types of kilns from pit and earth mound constructed with brick and stone to metal kilns and retorts. The main variation lies in the size of the kiln constructed for the purpose of churning out charcoal in sufficient volumes to sustain the supply of market needs. In less developed countries in Asia and Africa, pit and mound types continue to dominate the kilns used in all charcoal businesses because these types require not much initial capital costs by compromising manpower requirement and operation efficiency (Branch and Martiniello 2018: Girard 2002). Developed countries may employ more sophisticated kilns designed for the business production of charcoal efficiently and cost-effectively. These designs improve the nature of kiln itself, method of loading or charging, firing, and discharging. Whereas some business models use well-designed and constructed brick and stone kilns, others use metal-fabricated mobile or fixed kilns. The objective of all business kilns is to improve efficiency and system operation to achieve products with consistent high quality.

(iv) City-Type Kiln (Nonsmoking Type)

The siting of charcoal kilns depends on the nearness to the source of raw materials, consumers, or market. Another factor of great importance to affect the kiln siting is the impact of kiln operation on the surrounding environment. If a kiln is located near residential areas, the smoke and gases escaped from the kiln, which may be intolerable, are a nuisance to the community. Therefore, the quest to find smokeless kilns has been a primary objective of research and investigation until now. Several variations of these smokeless kilns have been produced, but most are of smaller and laboratory scales. Successful nuisance-free kilns have complicated issues to consider on pretreating the wood, firing procedure, and capturing the escaping gases. Because wood that contains much moisture and other volatiles tends to yield higher flue, pretreating the wood by drying tends to reduce the production of flue gas. Using electric heating results in near smokeless carbonization but makes the product more expensive in the end. Recapturing and treating the flue gases require intricate design and construction that makes the kiln more expensive to construct. Notwithstanding, a number of smokeless systems have been designed and tested and have been demonstrated to achieve various levels of successes. For example, the Sasukenei kiln shown in Fig. 3.12 follows the successful design of the Unbu kiln, and a smokeless kiln is developed by Kusakabe for Burnaby city in British Columbia, Canada (Jennifer 2010). There have been several models of smokeless kilns on the market lately, but their efficiencies have not been well documented.

Characteristics of different charcoal kiln types are summarized for community, commercial, and city as shown in Table 3.2.

(a)   **(b)**

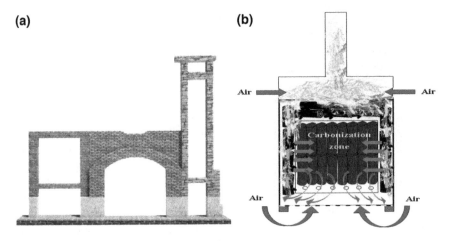

Fig. 3.12 (a) A model of Sasukenei kiln, (b) carbonization concept design

Table 3.2 Characteristics of different charcoal kiln types

Type	Size	Object	Wood Vinegar	Environment Impact
Community	Small	Used/small business	Yes	Yes
Commercial	Big	Business	Yes	Yes
City	Small	Used	No	No (small)

Box 3.1 Impact of Charcoal/Biochar Application to Community in Thailand

SIA (Social Impact Assessment) and SROI (Social Return on Investment) are the frameworks for measuring and evaluating the broader concept of value. The framework seeks to reduce inequality and environmental degradation as well as improve well-being by incorporating social, environmental, and economic costs and benefits. SROI is about value rather than money because money is simply a common unit widely accepted to convey value. SROI can help the project target appropriate resources for managing unexpected outcomes, both positive and negative.

This chapter presents the SROI in community charcoal production of Baan Phaisriaroon Farmer Group, Tha Luang Sub-district, Muang District, and Phichit Province in Thailand (Fig. 3.13). The methodology of this study was to establish scope, identify stakeholders, map outcomes, confirm outcomes and assign values, establish impact, and calculate the SROI. The techniques for collecting raw data included one-to-one interviews and meeting with a focus group. The impact value chain is presented in Table 3.3. When analyzing revenues and expenditures concerning the stakeholder groups, the life cycle assessment (LCA) principle is used to scope from raw material procurement, production, and transportation to disposal.

Fig. 3.13 Community training for charcoal briquette production

(continued)

Box 3.1 (continued)

Table 3.3 Impact of the value chain

Inputs	Activities	Outputs	Outcomes
1. Project investment for Baan Phaisriaroon Farmer Group	1.Training on installation and operation	1. Income of selling charcoal/biochar	1. Better life quality
2. Investment cost of the making charcoal/biochar system	2. Monitoring on the operation and maintenance	2. Savings in LPG purchasing	2. Safe environment
3. Knowledge about charcoal/biochar making transfer to the farmer group		3. Amount of LPG consumption	
4 Time spent engaged in activities of stakeholders		4. Level of CO_2 emissions from less LPG consumption	
		5. Transfer technology to the community	

Financial proxy for the monetization was collected from the stakeholders and related researchers and then calculated using the following Eq. 3.1:

$$\text{Present Value} = \frac{\text{Value of Impact in Year 1}}{(1+r)} + \frac{\text{Value of Impact in Year 2}}{(1+r)^2} + \cdots$$
$$+ \frac{\text{Value of Impact in Year } n}{(1+r)^n} \tag{3.1}$$

where

    r: Coupon rates of floating rate bonds from Bank of Thailand

    n: Project lifetime 1 year

$$\text{SROI ratio} = \frac{\text{Present Value}}{\text{Value of inputs}} \tag{3.2}$$

Sensitivity analysis was carried out to determine the value of environmental and social outcomes and returns that are sensitive to three factors: Case 1, cost of production; Case 2, selling price; and Case 3, reducing LPG costs (Table 3.4).

In the assessment, analyses of revenues and expenses were carried out for stakeholder groups and a participatory outcome chain. The system input is the

(continued)

Box 3.1 (continued)

Table 3.4 Investment and sensitivity analysis

Scenario	SROI	ROI	PB (month)	NPV (Baht)
Base case	3.71	3.70	3.24	1,471,353
Case 1: cost of production				
Cost increase by 70%	2.18	2.18	5.51	1,090,100
Costs increase by 50%	2.48	2.47	4.86	1,199,030
Costs increase by 30%	2.86	2.85	4.21	1,307,959
Case 2: biomass fuel price				
Selling price 10 baht/kg	2.66	2.64	4.54	895,353
Selling price 7 baht/kg	1.86	1.85	6.48	463,353
Selling price 5 baht/kg	1.33	1.32	9.08	175,353
Case 3: reduce the use of LPG, reduce the volume of charcoal sales	3.66	3.64	3.29	1,438,473

biomass technology, and the output is the annual quantity of charcoal produced, which has a heating value between 5814 and 6521 kcal/kg. The community development association is responsible for operating and maintaining the system with the outcome emphasizing the increased revenue for community members from the sale of the charcoal rods to improve the community quality of life while reducing waste pollution throughout the whole year. The environmental and social assessment of the community biochar production system investment showed 3.71 for SROI, 3.70 for ROI, and 1,471,353 Baht for NPV with a payback period of 3.24 months. In addition, maintaining the quality of biomass production is important. Biomass stack impacts the product sale prices and the investment revenue. The theory of change depicts that "If there is a promotion of biomass production, the farmers will have a better quality of life and environment." Based on this theory, long-term measures should be considered when demonstrating the effects or factors associated with the improved quality of life and environmental quality over time. SROI focuses on and emphasizes the need to measure the value from the bottom-up including the perspectives of different stakeholders. Hence, this is a novelty issue for new biomass projects beyond the scope of most previous studies about charcoal/biochar focusing on technical, economic, environmental, and policy issues. The SROI value of more than one may not represent high or acceptable benefits of the project or benefit-cost ratio. However, for completing this tool of assessment, other issues have to be integrated to assists project plans in achieving social sustainability.

3.2 System for Carbonizing Biomass from Agricultural Residue in Egypt

3.2.1 Challenges of Traditional Charcoal Production in Egypt

Charcoal or biochar is the remaining solid residue when wood, agro-industrial wastes, and other forms of biomass are carbonized under controlled conditions in a confined space such as a kiln or other traditional facilities. The carbonizing process that transforms biomass through slow pyrolysis takes place in four major stages with different temperature requirements in each stage as shown in Fig. 3.14 (Seboka 2009).

Making charcoal from alternative non-wood feedstocks usually involves a carbonization step followed by a briquetting step with a binder. Four main categories of small-scale and semi-industrial charcoal kilns can be identified: earthen kilns, brick kilns, metal kilns, and semi-industrial retorts. The highest and most consistent carbonization efficiency can be achieved by using semi-industrial retorts; but due to unavailable know-how and high investment costs, these retorts are often not affordable in Egypt. The biomass morphology can also limit the suitability of a certain type of carbonization kiln. To achieve higher conversion efficiencies and improve environmental performance, installing chimneys as well as facilities to recover tar and syngas recovery facilities is worth investigating. The traditional charcoal production in Egypt uses the earth mound technique (Fig. 3.15).

This type of technique has dominated the charcoal production in Egypt for a long time. The biomass is gathered, cut to size, and piled up on the ground to form a mound. The biomass pile or mound is then covered with earth that forms a gas-tight insulating barrier shell on the outside of the mound. Thus, the biomass carbonization can take place inside the earth barrier without air leakage so that the pile of biomass is fired and pyrolyzed without being burnt to ask. The pile of biomass is mostly sealed carefully by using fine soil, and a few air pockets are initially left open for steam and smoke to escape. As the pile emissions change color, the charcoal producer may seal some air pockets. After the carbonization process has ended and completed, the pile is opened or dug up to remove the charcoal when the

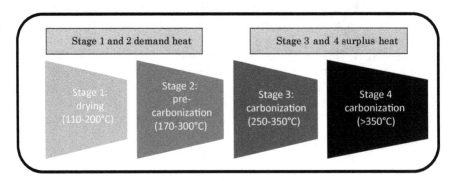

Fig. 3.14 The four stages of carbonization process and temperature requirements

Fig. 3.15 Traditional charcoal production in Egypt

temperature of the pile becomes lower. The conversion efficiency of this type of technique is typically about 10–15%.

The "black cloud" is a vernacular term that Egyptians have coined to refer to a recurring problem of air pollution due to the current use of the traditional charcoal production method. Hence, one significant contributor to air pollution in Egypt is the charcoal industry. About 7 million tons of wood residues from fruit tree are used as the feedstock for charcoal manufacturing in Egypt in more than 900 traditional open pile kilns. The emissions from these pits cannot be controlled so that many harmful pollutants including CH_4, SO_2, and NOx, tar, hydrocarbons gas, and volatile organic compounds are emitted to the atmosphere (El Safty and Siha 2013).

Charcoal is a national income for a number of families in Egypt; it is exported to some Arab and European countries thus bringing hard currency back to Egypt. The export quantity is about 30,000 tons charcoal per year according to the information issued by the government. However, issues on health hazards and environmental impact from time to time cause the crisis of Makamir (pile of biomass for producing charcoal) to loom over on the surface because these issues are caused by Makamir operations. Makamir is located in several governorates in Egypt; it is spread out in the governorates of Qalioubia, Gharbia, Mansoura, Alexandria, Beheira, and other governorates. The problem caused by this industry using Makamir to produce charcoal remains one of the most serious crises affecting responsible Egyptian government agencies although the export revenues is valuable to the nation. Therefore, the Minister of Environment has taken several measures to address this problem such as bringing the export of charcoal to an end and imposing fines to owners of the Makamirs that operate the obsolete kilns to cause adverse health and environmental impacts. The fines of about 5000 pounds may be as high as ten times the old fine. In the governorate of Qalioubia, specifically the village of Aghour Al-Kubra, most of the more than 100 Makamirs were in a state of stagnation and stopped producing charcoal. Then it became clear that the Makamirs were closed by the Ministry of Environment primarily because of the great damages caused by its use in charcoal production.

According to the Ministry of Environment statistics published in 2016, coal trade in Egypt is estimated to be about 11 million dollars. This value is likely to increase after the Ministry of Environment licenses charcoal producers and develops a working plan to regulate the industry. Compared with the level of exporting charcoal for other nations in the world, the charcoal trade is profitable and huge and has a great place in the Egyptian economy. A reasonable development of this industry will be of great importance to alleviating the pollution while reducing the unemployment rate and improving the economy.

3.2.2 Advanced Carbonization System in Egypt

The new mechanized metal charcoal kiln is expected to improve local air quality by reducing emission of greenhouse gases, especially carbon monoxide and carbon dioxide, and protect the local groundwater quality and land resources as well as vegetation and agricultural production from tar emissions. From the viewpoint of efficiency, this new metal kiln decreases charcoal productivity and quality although it takes less space to operate. On the other hand, its operation alleviates health hazards to and improves safety of charcoal producers and local residents. Unfortunately, the Ministry of Environment stipulated that owners of the Makamir have to adopt advanced models (iron kiln) to replace the traditional unit. However, this advanced kiln is not suitable to produce high-quality charcoal that conforms to the quality specifications of charcoal exported to Arab and European markets. Additionally, iron kiln as shown in Fig. 3.16 sometimes may explode due to improper operation.

Fig. 3.16 Charcoal production by using a steel kiln

Because of the abovementioned reasons, the charcoal producers in Egypt have not accepted this iron kiln. In the meantime, with the outbreak of the energy crisis, a new model of TH (thermal house) developed by El Sharkawi in a project funded by the Academy of Scientific Research and Technology (ASRT), Egypt, emerged in 2016. This model is capable of treating the gas emitted from the wood carbonization in addition to producing charcoal with quality in conformity with the export specifications and protecting the physical condition and the health of the operators. Most of the charcoal producers agree with TH that they are following with the same method of carbonization as mentioned before to operate this new TH.

The carbonization recycling system for agricultural residue biomass is a traditional technology practiced in Egypt because this method is developed by using more advanced industrial technology. Two systems have been developed: the first system that uses a thermal house technique with low emission of GHG to produce ordinary charcoal from the recycled residue wood of fruit trees and the second system that is capable of accepting various types and sizes of agricultural residue biomass. Other types of agricultural residues such as rice husk, rice straw, cotton stalk, palm tree residue, and trimming trees are recycled as biochar in the form of briquette charcoal to be used as energy source, slow-release fertilizer, active carbon, biochar hydrogel composite for soil to hold water, biochar urine composite, and wood vinegar and tar, among others. The carbonization process is continuously operated in this system to produce 500–750 kg biochar per day with self-sustained combustion. A similar system that is continuously operated as a screw reactor was also developed by ASRT, Egypt, in 2018. With an effective cascade arrangement, this system can increase the carbon content by fixing more carbon to the biochar effectively. By using this system, the so called "carbon-negative" effect in which carbon dioxide emissions are reduced in the carbonization process by returning the gas to the soil is expected to occur.

(i) Thermal Houses

A properly constructed and operated TH (thermal house) shown in Fig. 3.17 is without a doubt one of the most effective methods for charcoal production in Egypt nowadays. The experience of using this system over time has proved that it requires low capital investment and moderate labor costs to produce charcoal of surprisingly good yields and high quality. It is accepted by Egyptian producers for industrial and domestic users applying two TH designs with different capacities; both are capable of producing good result and are currently in extensive use throughout Egypt.

TH must comply with a number of important requirements to be successful. It must be simple to construct, relatively unaffected by thermal stresses during system heating up and cooling, and strong enough to withstand the mechanical stresses of loading and unloading in addition to being weatherproof over the carbonization cycle of about 30 days.

TH allows the entry of air at all times and during the cooling phase, but it is readily sealed hermetically to prevent entry of air. Using a heavyweight construction and fiberglass panels against fire and humidity, the system resists wind and erosion and provides good thermal insulation. The maximum temperature inside TH does not

Fig. 3.17 Thermal houses unit for charcoal production

exceeded 60 °C around the pile of wood as controlled by using a thermal control panel.

The advantages of TH are listed as follows:

1. It can be built in medium and large sizes.
2. It is built entirely of fiberglass panels locally made. It requires no steel except a few bars of flat steel to reinforce doors and the base of the dome.
3. It is robust and is not easily damaged either physically or thermally. It can be operated in open air to withstand sun, rain, or other ill weather effects without corrosion with a long useful life of about 8–10 years.
4. Electrical control panel of TH is relatively simple and equipped with temperature control.
5. The produced charcoal is suitable for all applications including household and metallurgical industry uses.

The major disadvantage of TH is that recovery or recycle of by-product tar or gas is not practiced in this system. This increases air pollution and slightly lowers the thermal efficiency. However, brick kilns have not been proved to be capable of recovering tar if the kiln uses more expensive and complex steel components.

(ii) **Carbonizing Products from Agricultural Waste**

About 10% of the 60 petagrams (Pg) of carbon fixed annually through photosynthesis worldwide ends up in agricultural residues (Dutta 2014). Approximately

30–35 million tons (2012) of different types of agricultural residues are generated in Egypt (NSWMP 2011), and approximately 12–15 million tons/year of agricultural waste is unused and disposed of by burning (El Essawy 2014). Burning unused agricultural waste in open fields, especially rice straw and rice husk, and trimming tree residues are a common practice by Egyptian farmers that contributes to the seasonal "black cloud" phenomenon. Despite the efforts put forward by the Ministry of Environment to discourage this practice, burning remains the most convenient way for farmers to dispose of waste because of high costs associated with collection, storage, and transportation of agricultural residues. However, there have been some important efforts to collect and process rice residues differently in order to reduce the negative impacts. The Egyptian Environmental Affairs Agency (EEAA) and the Ministry of Agriculture collected and handled 365,274 tons of the residues in 2011 (EEAA 2012).

The quantity of agricultural wastes is estimated to be about 35 million tons per year with 23 million tons of vegetable wastes in which 7 million tons are utilized as feed and 4 million tons as organic fertilizer. However, the remaining 12 million tons are left unused. In addition, only 3 of the 12 million tons per year of animal wastes are utilized with 9 million tons per year left unused. Thus, the total amount of unused agricultural wastes including plant residues and animal wastes is 21 million tons per year. These unused wastes lead to environmental pollution in addition to causing health hazards. On the other hand, the agricultural residue can be regarded as an enormous potential of untapped wealth. The waste of about 4.6 billion tons per year on average for the period of 2004–2012 can be considered as natural resources that may be turned into useful products with huge economic return. This will help advance the farm productivity especially for the farmers living on the newly developed farmland and desert (Bayoumi et al. 2014). Through a heat-induced chemical conversion process, such residues can be converted into biochar, which is a form of carbon that can be applied as a soil amendment, thereby providing long-term storage, or sequestration, of carbon in soil (Dutta 2014).

The crude gas generated from the wood pyrolysis process will be subject to different distillation steps to obtain combustible gas (CO, H_2, CH_4), tar, and other by-products (wood vinegar) that are to be used either for energy generation or agro-environmental conservation. Additionally, the whole production process is pollution-free.

The carbonization system does not need additional gas, oil, or any other fuels so the processing cost is quite low. The biochar unit can be used for only fine agricultural residues such as palm tree residues, corn stalk, or any source of organic materials generated from agricultural sectors except wood. There is only a small amount of flue gas emitted from the carbonization process. A screw reactor of the biochar production unit equipped with a corresponding presorting system according to the analysis of local waste components is used to maintain the optimal conditions for carrying out the carbonization operation. The thermochemical process converts biomass into more convenient products with higher values such as biochar, bio-oil, and syngas according to the source of material as shown in Fig. 3.18.

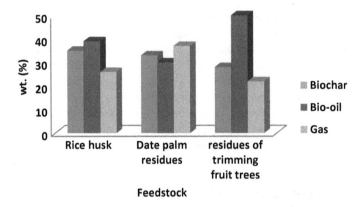

Fig. 3.18 Yields of the products obtained from the pyrolysis of the biomasses

3.2.3 Benefit of Developing Residual Biomass

Biochar has many applications in food security, renewable energy, and waste management that will help alleviate climate change. It is a carbon-rich charcoal produced through thermal pyrolysis of biomass under conditions of little or zero oxygen. The abundant feedstock that can be converted into biochar includes the carbon waste streams from agriculture residues, chicken residues, and other compostable wastes. All of these residues are low-value materials with limited uses but are costly to be disposed of properly. Once converted into biochar, these materials can be used as soil additives to enhance plant growth rates. Thus, this system provides an effective sink for sequestering atmospheric carbon dioxide. Other benefits are listed as follows:

- Less risk of reduced crop yield during dry seasons
- Reducing the need for chemical fertilizers
- Retaining nitrogen and other elements in soil that also reduce emissions
- Filtering out contaminants from shallow soil water
- Removing heavy metals and acids from abandoned mine ponds
- Binding toxins and preventing their leaching into surface and groundwater

Currently, a variety of biochar derivatives with different properties depending on the feedstock, pyrolysis condition, residence time, and additives added have been produced. Standardization and classification of the various types are required if the product is marketed for public use. Biochar can be used as a relatively low-cost agent through applying inexpensive technology to mitigate climate change and enrich soil in large scale with timely results. Also, this method provides an alternative measure for small municipalities to recycle their increasing quantity of sewage sludge and other organic wastes in a sustainable manner.

3.2.3.1   Why Biochar Is Beneficial Material for Soil?

The physical and chemical nature of biochar brings about a unique capability for biochar to attract and hold moisture, nutrients, and even agrochemicals; the latter are nutrients that are difficult to be held in soil like nitrogen and phosphorous. Nitrogen tends to be carried out of regular soils by surface runoff to upset the ecosystem balance in streams and riparian areas. With its immense surface area and complex pore structure with 500 m^2 surface area per gram, biochar provides a secure habitat for microorganisms and fungi to grow (https://www.tecnica.com.au/). Certain fungi form a symbiotic relationship with plant root fibers, thus allowing greater nutrient uptake by plants.

3.2.3.2   Why Biochar Is Beneficial for Plant Production?

When added to soil, biochar improves plant growth, enhances crop yields, and increases food production sustainability in soils with depleted nutrients, limited organic resources, and insufficient water and/or access to agrochemical fertilizers. Not all soils react similarly to the addition of biochar, and it may frequently take up to a year to realize the beneficial results. Many studies show that biochar can increase crop yields of poor soils with low carbon content up to four times. Research presented at a recent American Chemical Society annual meeting suggests that biochar plus chemical fertilizer enhances growth of winter wheat and several vegetables by 25–50% as compared with chemical fertilization alone. The experiment conducted by the Soil Science Society of America reveals that biochar supplemented with fertilizer outperforms fertilizer alone by 60% (http://biochar-us.org/soil-water-benefits-biochar).

3.2.4   Value-Added Products Derived from Biochar

Biochar is a by-product derived from low-value agricultural residues through the pyrolysis process (Fig. 3.19). It is an intermediate material for many of the products used for agriculture, environment, and industry. So far, the benefits of all biochar products have not been fully developed yet such as for slow-release fertilizer, super-retention of irrigation water, water purification, pesticide and insecticide application, and charcoal for energy.

(i) **Biochar Ammonium Phosphate**

An innovative novel BAP (biochar-ammonium phosphate), which is an uncoated slow-release fertilizer, is used for fertilization of sandy soil. It is based on the reaction between biochar and phosphoric acid or ammonia gas and can be considered as

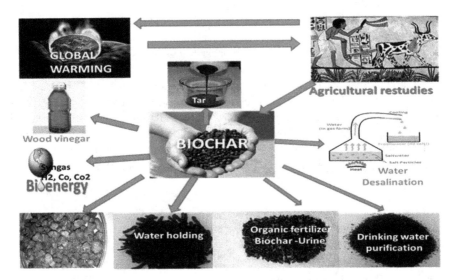

Fig. 3.19 The by-products of pyrolysis

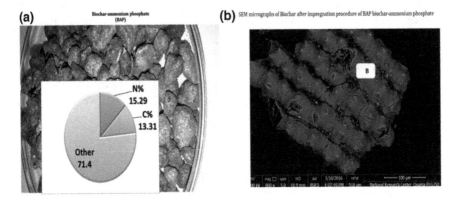

Fig. 3.20 Biochar ammonium phosphate (BAP)

a nano-store material for nutrients not only from the surface but also from inside the pores on the surface. Application of BAP fertilizer to sandy soil has another benefit of supporting the sandy soil with organic carbon. The cellular microstructure of the biochar as shown by the SEM micrograph illustrated in Fig. 3.20 reveals multiple hollow channels of different diameters generated from tracheid cells. Also, the crystalline structures of BAP can be seen in Fig. 3.20 (b) in the center and the outer surface of the biochar particles in addition to the cell pores, which are uniformly loaded with fertilizer after the reaction process. Figure 3.20 (a) shows the final shape of the finished BAP.

(ii) **Biochar Hydrogel Composite**

3.2.4.1 What's Biochar Hydrogel Composite?

Biochar hydrogel composite (BHC) is the biochar activated with polymer hydrogel. It is classified as not only a superabsorbent polymer but also a super source of organic carbon. BHC is viscoelastic with loosely crosslink and hydrophilic three-dimensional networks of versatile polymer chains inside and around the pores. This composite enables BHC to absorb and retain the quantity of water or biological fluids a hundred times its own weight within a short period of time under stressful conditions. Hence, it is considered as a soil conditioner to hold water in the soil and an agent to increase crop yield.

Three types of BHC are as shown in Fig. 3.21:

1. Biochar activated with 20% hydrogel
2. Biochar activated with 30% hydrogel
3. Biochar activated with 50% hydrogel

3.2.4.2 Mode of Action for BHC

When BHC is mixed with the soil around the plant, it undergoes hydration to form an associated amorphous gelatin-like mass with excellent absorption and desorption capabilities for an extended period of time.

The BHC particle is also considered as a miniature water reservoir and organic carbon stock in the soil. Water will be detached from these reservoirs by plant root through osmotic pressure difference. After water is released to the plant, the respective BHC volume is reduced thus creating free pore volume or space for air and water infiltration and storage. Hence BHC polymer plays the role as a slow-release source of water and dissolved fertilizers in the soil.

Fig. 3.21 Biochar hydrogel composite (BHC)

Water conservation by BHC creates an effective buffer against short-run drought tension and loss reduction in institution phase mainly in regions depending primarily on the rain for plant growth especially when rainfall is insufficient and intermittent. Once BHC is mixed into soil, it preserves water and nutrients in vast quantities a hundred times of its original weight; the water and nutrients are then released as required by the plant; therefore, plant growth can be enhanced with limited water supply.

Benefits of BHC include (a) enhancing soil water-holding capacity, (b) increasing soil permeability, (c) improving water retention of different soil types, (d) increasing the water use efficiency, (e) increasing soil organic carbon, (f) benefiting especially sandy soil, (g) extending irrigation intervals due to increasing time to reach a permanent wilting point, (h) minimizing soil erosion and surface water runoff, (i) enhancing soil penetration and infiltration, (j) decreasing soil compaction tendency and improving soil drainage, (k) supporting crop growth performance under reduced irrigation conditions, (l) enhancing nutrient retention as a result of solute release from BHC particles and delaying the dissolution of fertilizers, and (m) reducing cost as compared with the use of hydrogel alone.

(iii) **Charcoal Briquettes**

Biochar derived from agricultural residues may be not only used for soil amendment but also processed into briquette (Fig. 3.22). Briquette biochar is used as an

Fig. 3.22 Briquette biochar processing

abundant clean fuel in cook stoves that is not affected by the availability of harvested wood used to make charcoal. How is using the biochar briquette in question different from using other types of biomass fuel? Green heat briquette has a higher bulk density and lower moisture content as compared with conventional biomass fuel. Replacing charcoal with biochar briquette reduces significant amount of smoke during ignition with longer burning time thus making the fuel last longer and saving more money for the customers. The uniform shape and strength help maintain relatively stable temperatures that lead to less char wastage. This kind of briquette is odorless because it contains minimal quantity of evaporative substances. Burning charcoal briquettes is also safer because it produces no sparks like wood charcoal. Our alternative clean fuel can replace wood and charcoal for use in all conventional biomass stoves.

Additionally, biochar is two to four times less expensive than charcoal so that users can achieve significant savings in fuel costs. These savings enable them to buy better stoves that further lower their energy consumption and the associated costs. A portion of the revenue from the project is reinvested in reforestation projects to help alleviate the emission of greenhouse gases.

Social, environmental, and economic impacts of using biochar as an alternative fuel to the traditional wood charcoal are quite significant. The initiative of using smokeless and odorless biochar helps protect people's health. Users also save money on fuel costs thus assisting them in meeting their own financial needs and those of their families as well that improves their living conditions. As the environmental impacts are concerned, the initiative alleviates deforestation by providing an alternative to reduce deforestation. Part of the benefit of using biochar briquette is actively reinvesting the profit in reafforestation in addition to raising environmental awareness in the community. This kind of activity also creates new jobs for the youth and women in the region where the agricultural residues are accumulated. In addition, profits earned on the sale of biochar arouse the initiative to make further investments and open up new sources of income for the beneficiaries.

3.2.5 Biochar Application to Agricultural Land in Egypt

The Egyptian government (Ministry of Agriculture) has reclaimed and added about 3 million acres of new land to the cultivated area recently. The newly added sandy soil is incapable of retaining irrigation water and nutrients because it contains less amount of organic matter. Thus, farmers apply organic matter to improve the soil fertility. New virgin lands in the Delta of Egypt are characterized by the absence of many diseases and contaminants that may be suffered by the old farmland in the same region. Many of the plant root diseases, such as nematodes and fusarium, cause numerous concerns for farmers on yield loss, excessive application of pesticides, and the resulting high production costs and low quality of the crop. In addition, organic fertilizers derived from animal residues or mostly immature organic compost are used to improve the soil. Using this practice, farmers directly infect their newly developed land with various diseases. Problems caused by pollutants such as heavy elements and hazardous chemicals pose a great threat to the continuity of purity for the newly developed agricultural land and therefore its capability to produce agricultural products of high quality for either domestic consumption or export. Hence, the benefit of cultivating new land will be lost because the crop plants growing on their land suffer from the spread of root diseases and chemical pollutants that are brought by immature organic matter. Applying alternative organic materials such as biochar, which is free from pollutants, as an alternative organic material from traditional organic residues to such barren land is of great importance to enhance the crop.

(Note: The production of compost in an appropriate and scientific manner is not practiced in Egypt due to the lack of economic incentive.)

References

Aabeyir R, Adu-Bredu S, Agyare WA, Weir MJC (2016) Empirical evidence of the impact of commercial charcoal production on Woodland in the Forest-Savannah transition zone, Ghana. Energy Sustain Dev 33:84–95. https://doi.org/10.1016/j.esd.2016.03.005

Antal MJ, Grønli M (2003) The art, science, and technology of charcoal production. Ind Eng Chem Res 42(8):1619–1640. https://doi.org/10.1021/ie0207919

Antal MJ Jr, Mok WS, Varhegyi G, Szekely T (1990) Review of methods for improving the yield of charcoal from biomass. Energy Fuel 4(3):221–225

Bayoumi HHA, El Gebaly MR, Abdul Ghani SS, Hussein YMM (2014) An economic study of recycling agricultural wastes in Egypt. Middle East J Agric Res 3(3):592–608

Branch A, Martiniello G (2018) Charcoal power: the political violence of non-fossil fuel in Uganda. Geoforum. https://doi.org/10.1016/j.geoforum.2018.09.012

Byrne CE, Nagle DC (1997) Carbonization of wood for advanced materials applications. Carbon 35(2):259–266. https://doi.org/10.1016/S0008-6223(96)00136-4

Castro JP, Nobre JRC, Bianchi ML, Trugilho PF, Napoli A, Chiou B-S et al (2018) Activated carbons prepared by physical activation from different pretreatments of amazon piassava fibers. J Nat Fibers:1–16. https://doi.org/10.1080/15440478.2018.1442280

Chan WCR, Kelbon M, Krieger-Brockett B (1988) Single-particle biomass pyrolysis: correlations of reaction products with process conditions. Ind Eng Chem Res 27(12):2261–2275

Chiang P-C, Pan S-Y (2017) Analytical methods for carbonation material carbon dioxide mineralization and utilization. Springer, Dordrecht, pp 97–126

Chrzazvez J, Théry-Parisot I, Fiorucci G, Terral J-F, Thibaut B (2014) Impact of post-depositional processes on charcoal fragmentation and archaeobotanical implications: experimental approach combining charcoal analysis and biomechanics. J Archaeol Sci 44:30–42

Das K, Hiloidhari M, Baruah D, Nonhebel S (2018) Impact of time expenditure on household preferences for cooking fuels. Energy 151:309–316

Deforce K, Boeren I, Adriaenssens S, Bastiaens J, De Keersmaeker L, Haneca K et al (2013) Selective woodland exploitation for charcoal production. A detailed analysis of charcoal kiln remains (ca. 1300–1900 AD) from Zoersel (northern Belgium). J Archaeol Sci 40(1):681–689. https://doi.org/10.1016/j.jas.2012.07.009

Dokoohaki H, Miguez FE, Laird D, Horton R, Basso AS (2017) Assessing the biochar effects on selected physical properties of a sandy soil: an analytical approach. Commun Soil Sci Plant Anal 48(12):1387–1398

Dokoohaki H, Miguez FE, Archontoulis S, Laird D (2018) Integrating models and data to investigate the effect of biochar on soil hydrological properties. In: The promise of biochar: From lab experiment to national scale impacts, vol 30. Iowa State University, Iowa

Dorner-Reisel A, Yoseph Y, Matner V, Klemm V, Svoboda S (2017) Investigation of carbonation of wheat stems from Central Europe during slow pyrolysis at different temperatures. J Environ Agric Res IJOEAR 3(1):30–39

Dutta B (2014) Development and optimization of pyrolysis biochar production systems towards advanced carbon management. Thesis of Doctor of Philosophy. McGill University, Department of Bioresource Engineering

EEAA (2012) Statistics of environment and energy in Egypt. Egyptian Environmental Affairs Agency (EEAA), Cairo

El Essawy M (2014) Ministry of environment monitors agricultural waste uses, The Cairo Post, 30th January, 2014. Retrieved from: thecairopost.youm7.com/news/82684/business/ministry-of-environment-monitors-agricultural-waste-uses

El Safty A, Siha M (2013) Environmental and health impact of coal use for energy production. Egyptian J Occ Med 37(2):181–194

Gay-des-Combes JM, Robroek BJ, Hervé D, Guillaume T, Pistocchi C, Mills RT, Buttler A (2017) Slash-and-burn agriculture and tropical cyclone activity in Madagascar: implication for soil fertility dynamics and corn performance. Agric Ecosyst Environ 239:207–218

Girard P (2002) Charcoal production and use in Africa: what future? Unasylva 53(4):30–35

Goulart AC, Macario KD, Scheel-Ybert R, Alves EQ, Bachelet C, Pereira BB et al (2017) Charcoal chronology of the Amazon forest: A record of biodiversity preserved by ancient fires. Quat Geochronol 41:180–186. https://doi.org/10.1016/j.quageo.2017.04.005

Habte Y, Hector D (2017) Business model for Black Pellets production in Sweden

Hagemann N, Spokas K, Schmidt H-P, Kägi R, Böhler M, Bucheli T (2018) Activated carbon, biochar and charcoal: linkages and synergies across pyrogenic carbon's ABCs. Water 10(2):182

Jennifer H (2010) The Sasukenei smokeless kiln: A wood kiln that produces little smoke and great results. ceramic kilns. Retrieved from https://ceramicartsnetwork.org/daily/clay-tools/ceramic-kilns/the-sasukenei-smokeless-kiln-a-wood-kiln-that-produces-little-smoke-and-great-results/

Lehmann J, Joseph S (2015) Biochar for environmental management: science, technology and implementation. Routledge, London

Lennox S (2017) Towards an understanding of middle stone age wood uses 58 000 years ago in KwaZulu-Natal: charcoal analysis from two Sibudu occupation layers, BYA2i and SPCA. S Afr J Bot 109:345

Maliwan Haruthaithanasan PD (2003) Charcoal production and utilization in Thailand. Go Green. Retrieved from https://www.oeaw.ac.at/forebiom/WS1lectures/Plenary_Haruthaithanasan.pdf

Maroušek J, Vochozka M, Plachý J, Žák J (2017) Glory and misery of biochar. Clean Techn Environ Policy 19(2):311–317

Maruyama T, Oliveira L, Britto A, Nisgoski S (2018) Automatic classification of native wood charcoal. Eco Inform 46:1

Monteiro PD, Zapata L, Bicho N (2017) Fuel uses in Cabeco da Amoreira shellmidden: an insight from charcoal analyses. Quat Int 431:27–38

Mulenga BP, Hadunka P, Richardson RB (2017) Rural households' participation in charcoal production in Zambia: does agricultural productivity play a role? J For Econ 26:56–62. https://doi.org/10.1016/j.jfe.2017.01.001

Ngatia K, Ngatia M, Ngatia J, Ng'oriareng P, Simanto O, Oduor N (2016) available charcoal production technologies in Kenya. UNDP, New York

NSWMP (National solid waste management programme) Egypt, December 22, 2011. http://www.eeaa.gov.eg/portals/0/eeaaReports/NSWMP/1_P0122721_NSWMP_Main%20Report_December2011.pdf http://biochar-us.org/soil-water-benefits-biochar. Soil & Water Benefits of Biochar

Pattnaik D, Kumar S, Bhuyan S, Mishra S (2018) Effect of carbonization temperatures on biochar formation of bamboo leaves. Paper presented at the IOP Conference Series: Materials Science and Engineering

Pereira BLC, Carneiro A d CO, Carvalho AMML, Colodette JL, Oliveira AC, Fontes MPF (2013) Influence of chemical composition of Eucalyptus wood on gravimetric yield and charcoal properties. Bio Res 8(3):4574–4592

Pereira EG, Martins MA, Pecenka R, Carneiro A d CO (2017) Pyrolysis gases burners: sustainability for integrated production of charcoal, heat and electricity. Renew Sust Energ Rev 75:592–600. https://doi.org/10.1016/j.rser.2016.11.028

Platt D, Workman M, Hall S (2018) A novel approach to assessing the commercial opportunities for greenhouse gas removal technology value chains: developing the case for a negative emissions credit in the UK. J Clean Prod 203:1003–1018

Puentes-Rodriguez Y, Torssonen P, Ramcilovik-Suominen S, Pitkänen S (2017) Fuelwood value chain analysis in Cassou and Ouagadougou, Burkina Faso: from production to consumption. Energy Sustain Dev 41:14–23

Raab A, Takla M, Raab T, Nicolay A, Schneider A, Rösler H et al (2015) Pre-industrial charcoal production in lower Lusatia (Brandenburg, Germany): detection and evaluation of a large charcoal-burning field by combining archaeological studies, GIS-based analyses of shaded-relief maps and dendrochronological age determination. Quat Int 367:111–122. https://doi.org/10.1016/j.quaint.2014.09.041

Richards T (2016) Biochar production opportunities for South East Asia

Rousset P, Aguiar C, Labbé N, Commandré J-M (2011) Enhancing the combustible properties of bamboo by torrefaction. Bioresour Technol 102(17):8225–8231

Sara ZS, Shah T (2018) Residual effect of biochar on soil properties and yield of maize (Zea mays L.) under different cropping systems. Open J Soil Sci 8(01):16

Seboka Y (2009) Charcoal production: opportunities and barriers for improving efficiency and sustainability. Bio-carbon Opportunities in Eastern & Southern Africa. UNDP, New York

Smith HE, Hudson MD, Schreckenberg K (2017) Livelihood diversification: the role of charcoal production in southern Malawi. Energy Sustain Dev 36:22–36. https://doi.org/10.1016/j.esd.2016.10.001

Svedelius G (1875) Hand-book for charcoal burners. Wiley, New York

Tang MM, Bacon R (1964) Carbonization of cellulose fibers – I. low temperature pyrolysis. Carbon 2(3):211–220

Tintner J, Preimesberger C, Pfeifer C, Soldo D, Ottner F, Wriessnig K et al (2018) Impact of pyrolysis temperature on charcoal characteristics. Ind Eng Chem Res 57:15613

Villota EM, Lei H, Qian M, Yang Z, Villota SMA, Zhang Y, Yadavalli G (2017) Optimizing microwave-assisted pyrolysis of phosphoric acid-activated biomass: impact of concentration on heating rate and carbonization time. ACS Sustain Chem Eng 6(1):1318–1326

Woolf D, Lehmann J, Joseph S, Campbell C, Christo FC, Angenent LT (2017) An open-source biomass pyrolysis reactor. Biofuels Bioprod Biorefin 11(6):945–954

Wu H, Lai C, Zeng G, Liang J, Chen J, Xu J et al (2017) The interactions of composting and biochar and their implications for soil amendment and pollution remediation: a review. Crit Rev Biotechnol 37(6):754–764

Xie X, Goodell B, Daniel G, Qian Y, Jellison J, Peterson M (2009) Carbonization of wood and nanostructures formed from the cell wall. Int Biodeterior Biodegradation 63(7):933–935. https://doi.org/10.1016/j.ibiod.2009.06.011

Chapter 4
Nutrient Recovery from Wasted Biomass Using Microbial Electrochemical Technologies

Hiroyuki Kashima

Abstract Human beings have developed technologies to effectively supply tremendous amount of nutrients, nitrogen and phosphorus, to the modern agriculture, and these synthetic fertilizers are used to grow crops that feed much of the global population. However, such anthropogenic activities that largely skew the global nutrient cycles are problematic in the light of sustainability. This chapter discusses problems of modern nutrient cycles associated with agricultural activities and then introduces technologies to recover those nutrients from wasted biomass that are essential to construct the anthropogenic nitrogen and phosphorus cycles in a sustainable form.

The production of tremendous amount of synthetic fertilizers that drive the modern agriculture is heavily dependent on energy and limited resources. The current synthetic nutrient supply for either nitrogen that relies on fossil energy to fix the atmospheric nitrogen gas and phosphorous that depends on mining a limited source has obvious risks in sustainability. Another issue is the negative environmental impacts associated with discharge of excess nutrients from agricultural fields and urban systems. The discharge of nitrogen and phosphorus into aquatic systems often results in eutrophication that is fatal to the ecosystem. Recovering nitrogen and phosphorus from materials currently wasted and using recovered nutrients as fertilizer are essential to construct the anthropogenic nitrogen and phosphorus cycles in sustainable form. Wastewaters discharged from urban and agricultural systems are the premier source of those recoverable nutrients because they are relatively easy to recover.

As a novel promising technology for nutrient recovery from wasted biomass, the latter part of the chapter introduces principles of microbial electrochemical technologies (METs) and their applications for nutrient recovery processes. METs are the hybrid technologies integrating microbial metabolism and electrochemical reactions for exploiting the capabilities of microbes to catalyze various electrochemical reactions to achieve simultaneous wastewater treatment and

H. Kashima (✉)
Super-cutting-edge Grand and Advanced Research (SUGAR) Program, Japan Agency for
Marine-Earth Science and Technology (JAMSTEC), Yokosuka, Japan
e-mail: kashimah@jamstec.go.jp

© Springer Nature Singapore Pte Ltd. 2020
S. Tojo (ed.), *Recycle Based Organic Agriculture in a City*,
https://doi.org/10.1007/978-981-32-9872-9_4

production of energy or chemicals. Some recent MET designs address nutrient removal/recovery so that this process will provide a novel nutrient management strategy for treating wastewater streams. More importantly, other studies on METs process have been demonstrated to be a potential alternative energy-efficient nutrient recovery process.

4.1 Nutrient Recovery from Wasted Biomass

4.1.1 Significance of Nutrient Cycles Associated with Agricultural Activities

Nitrogen (N) and phosphorus (P) are essential elements for the life on this planet. Human beings rely largely on supplying plant macronutrients containing N and P as fertilizer into agricultural systems to secure the supply of food and other important commodities. Historically, crops were produced by naturally available nutrients in the soil and the artificial input of locally available organic fertilizers such as manure and plant residues. By contrast, the modern agriculture is driven by the supply of tremendous amount of synthetic fertilizers. For example, the use of nitrogen fertilizer increased about eight times from 1960 to 2000 (Fixen and West 2002), and much of the fertilizer is produced by using the industrial nitrogen fixation method known as the Haber-Bosch process. The Haber-Bosch nitrogen fixation and the agricultural nitrogen fixation, which are about 1.6 times greater than the natural terrestrial nitrogen fixation, account for about 40% of the total nitrogen fixation on this planet (Canfield et al. 2010). With regard to phosphorus, 90% of the global demand is for food production, and around 15 Mt phosphorus of the fertilizer is produced per year by mining phosphate rock (Cordell et al. 2009). Today, synthetic fertilizers are used to grow crops that feed much of the global population.

However, human society is facing serious problems associated with modern agricultural systems, which consumes a considerable amount of synthetic fertilizers in agricultural fields. The first major problem is on the sustainable supply of those nutrient resources. Virtually unlimited supply of nitrogen is reserved in the atmosphere in the form of nitrogen gas, but only several types of prokaryotes that have the capability of nitrogen fixation with nitrogenase can utilize this chemically stable atmospheric nitrogen gas as the nutrient. Modern industries use energy-intensive physiochemical processes to convert atmospheric nitrogen gas into reactive nitrogen species such as ammonia that is readily available for plants to use. For example, the Haber-Bosch process as delineated by Eq. 4.1 consumes a considerable percentage of the world's natural gas production to satisfy the energy demand for producing nitrogen fertilizer (McCarty et al. 2011).

$$N_2(g) + 3H_2(g) \rightarrow 3NH_3(g) \quad \Delta H = -92.4 \, kJ/mol \qquad (4.1)$$

On the other hand, the phosphorus fertilizer is solely acquired from mining mineral deposits, and the source will be depleted if no other alternatives are developed. The global peak of phosphorus production is expected to be around the year of 2030, and the current global phosphate rock reserves may be depleted in 50–100 years (Cordell et al. 2009). As discussed in above paragraphs, the current synthetic nutrient supply for either nitrogen that relies on fossil energy to produce or phosphorous that depends on mining a limited source has obvious risks in sustainability.

Another issue is the adverse environmental impacts associated with discharge of excess nutrients from agricultural fields and urban systems. Because nitrogen and phosphorus are limiting nutrients in surface water bodies, their discharge into aquatic systems often results in eutrophication that induces overgrowth of phototrophic organisms such as algal bloom followed by hypoxia that is fatal to the ecosystem. Substantial amounts of nitrogen and phosphorus contained in fertilizers are applied to agricultural fields; some are incorporated into crops that are consumed by domestic animals and ultimately by human being. During this process, the incorporated nutrients are eventually discharged in wastes such as manure and sewage. On the other hand, a major portion of applied nutrients is not utilized by crops; they are lost in agricultural surface runoff. Studies conducted in the mid-1990s estimated that 40 Mt. of N out of 170 Mt. N of synthetic nitrogen fertilizer applied to the world's crop fields per year was either emitted into the atmosphere or discharged into adjacent water systems (Smil 2002). Also, some soil microorganisms transform a fraction of the field-applied reactive nitrogen fertilizer into gaseous nitrogen compounds that escape into the atmosphere. About 25% of nitrous oxide, which is about 300 times stronger in greenhouse effect than carbon dioxide (Mosier et al. 1998), as well as the dominant ozone-depleting substance (Ravishankara et al. 2009), are discharged from agricultural sectors. As the phosphorus is concerned, about 46% of mined phosphorus fertilizer is lost into freshwater systems through soil erosion and agricultural surface runoff. Additionally, a significant quantity of phosphorus that is about 40% of the mined phosphorus fertilizer is contained in animal wastes; it is discharged into surface water systems from pastures because of non-stringent surface runoff water quality regulations (Rittmann et al. 2011). As a result, the P content of terrestrial and freshwater ecosystems has increased at least 75% greater than pre-industrial levels (Bennett et al. 2001). Collectively, agricultural runoff and insufficient nutrient removal in manure/sewage treatment systems lead to extensive eutrophication of fresh surface water systems and coastal waters resulting in issues such as loss of biodiversity, fish kill, and deterioration of the water quality. Hence, proper management of nutrients discharged from agricultural fields and urban systems is necessary to alleviate the discharge of harmful pollutants (Fig. 4.1).

Sustainable supply of essential nutrients to agricultural systems while mitigating problems associated with discharge of those nutrients into the environment is a grand challenge to human beings for this century. The advancement of nutrient removal processes in wastewater treatment during the latter half of last century has eliminated the discharge of nutrients into local water systems at least under some situations. For example, wastewater treatment plants employ the strategy of removing dissolved reactive nitrogen compounds into gaseous phase to be emitted from

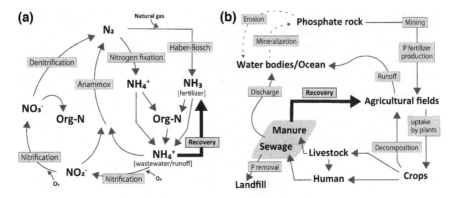

Fig. 4.1 Modern nutrient cycles under influence of human activities. (**a**) Schematic diagram of the modern nitrogen cycle. (**b**) Schematic diagram of the modern phosphorus cycle. (Diagrams was made based on the information of Nancharaiah et al. 2016)

the aqueous phase into the atmosphere. The most widely employed process for treating municipal and industrial wastewaters is the two-step nitrification/denitrification process. During the first nitrification step, the soluble ammonia is oxidized to nitrate by nitrifying microorganisms (Eq. 4.2). Subsequently, nitrate is reduced to nitrogen gas by denitrifying microorganisms with the amendment of electron donors in the second denitrification step (Eq. 4.3).

$$NH_4^+ + 2O_2 \rightarrow NO_3^- + H_2O + 2H^+ \tag{4.2}$$

$$2NO_3^- + 12H^+ + 10e^- \rightarrow N_2 + 6H_2O \tag{4.3}$$

This biological nitrogen removal scheme is effective in reducing reactive nitrogen compounds in the various wastewaters below discharge limits. However, this process is costly and energy intensive. Supplying oxygen to achieve nitrification by aeration requires considerable operational cost and electricity (Ledezma et al. 2015). Further, additional electron donors are required for the subsequent denitrification step. In recent years, novel energy-saving and cost-efficient nitrogen removal strategies have been developed. For example, anaerobic ammonium oxidation (anammox), microbial ammonia oxidation, coupled with nitrite reduction discovered in the 1990s (Jetten et al. 1998, Strous et al. 1999) has been widely studied as an energy-efficient nitrogen removal process without aeration (Lackner et al. 2014). Microbial electrochemical technologies (METs) to be discussed in the following sections have also been developed with integrated nitrogen removal processes such as biocathodic denitrification (Clauwaert et al. 2007) as an energy-offsetting process. However, current processes of nitrogen removal from waste streams are costly and energy intensive. Although the worldwide implementation of novel nitrogen removal processes has greatly reduced the energy consumption, these processes depend on converting aquatic reactive nitrogen into nitrogen gas for its removal from water. For supplying fertilizer to agricultural application, however, the industrial nitrogen fixation process is needed to convert nitrogen gas into reactive nitrogen. The modern anthropogenic nitrogen cycle driven by the combination of the Haber-Bosch nitrogen fixation process and the denitrification nitrogen removal processes is problematic in light of sustainability.

Processes of phosphorus removal from waste streams have also been developed. Phosphates contained in wastewaters are in organic suspended solid and dissolved forms. Solid organic phosphorus can be removed from waterbody by conventional physical operations such as sedimentation and rapid sand filtration. Effective removal of dissolved phosphate requires the use of advanced treatment operations such as membrane filtrations including reverse osmosis systems and membrane bioreactors. The level of phosphate contained in the treated water is relatively lower than that specified by current regulations, but these operations are too expensive to be implemented in most waste treatment facilities. Therefore, the major commercial phosphate removal process currently implemented is chemical precipitation with the addition of metallic coagulants such as ferrous or ferric salt, alum $(Al_2(SO_4)_3 \cdot 8H_2O)$, or lime $(Ca(OH)_2)$ (De-Bashan and Bashan 2004). The chemical precipitation method successfully removes water-borne pollutants and dissolved phosphorus from the treated water. But recovering the phosphorus retained in the resulting sludge for reuse as fertilizer is impractical. This continuous loss of phosphorus through wasting the phosphate-containing chemical sludge makes anthropogenic phosphorus use as one-way flow and problematic with respect to sustainable supply of the essential resource.

As discussed in above paragraphs, current anthropogenic cycles of the important nutrients such as nitrogen and phosphorus have issues with the sustainable supply of the nutrients in addition to their uncontrolled discharge causing adverse environment impacts. To overcome those challenges, we need to close the anthropogenic N and P cycles in sustainable form. The key would be recovering nutrients contained in the organic wastes as resource of fertilizer to be applied to agricultural systems. Therefore, developing the agricultural systems based on the recovered nutrients from organic wastes by using new nutrient recovery technologies is required to replace synthetic fertilizer-based agriculture and nutrient removal processes used in modern wastewater treatment systems. More discussions on cycled nutrients-based agricultural systems are presented in other chapters of this book. The following sections of this chapter focus on available technologies for recovering nutrient from organic wastes, especially from wastewaters.

4.1.2 Nutrient Recovery Technologies from Waste Streams

Recovering nitrogen and phosphorus from materials currently wasted or converted into unrecoverable forms and using recovered nutrients as fertilizer are essential to construct the anthropogenic nitrogen and phosphorus cycles in sustainable form. Wastewaters discharged from urban and agricultural systems are the premier source of those recoverable nutrients because they are relatively easy to recover. Wastewaters and wasted biomass with high water content derived from urban and agricultural systems reserve a major portion of the nutrients consumed and then wasted by human and domestic animals. For example, the amount of phosphorus contained in domestic wastewaters and sludge resulting from wastewater treatment constitutes most of the phosphorus consumed by human that accounts to 15% of the total mined phosphorus input (Rittmann et al. 2011). Some fraction of the phosphorus in animal waste, which amounts to about 40% of the total mined P, is treated as agricultural wastewaters as well. In addition, a considerable amount of domestic wastewater is systematically collected and treated in urban infrastructures such as centralized wastewater systems. This situation is exceptionally favorable to reusing the wastewater nutrient as feedstock because collecting sparsely distributed wasted biomass is one of the major challenges for wasted biomass refinery. Furthermore, the cost for recovering nutrient from wastewaters is alleviated because traditionally those wastewaters are treated with cost-intensive conventional nutrient removal processes for meeting legal discharge regulations. For instance, conventional biological nitrification with extensive aeration for nitrogen removal accounts for 60% of operational cost of treating wastewater (Nancharaiah et al. 2016).

However, nutrient recovery from wastewaters is not well accomplished particularly in urban areas. For example, less than 30% of human waste in urban areas in China and on average about 50% of urban sewage in European countries were recycled to agricultural systems in the 1990s (Liu et al. 2008). Globally, it was estimated that 20% of human wastes in urban area are recycled, while 70% are recycled in rural areas (Liu et al. 2008). Only a small fraction of nutrients is estimated to be recovered from wastewater treatment processes through wastewater reuse as well as reclamation and application of the recovered biosolids (Mihelcic et al. 2011). Most sewage solids are currently disposed of using landfills, and only about 1% is recycled to soil (Rittmann et al. 2011). Wastewater reuse and reclamation in agricultural sectors are difficult especially for urban areas, which are remote from agricultural fields. The agriculture in urban areas to be discussed in this book shows the potential to mitigate this hurdle by reducing physical distance between the source and the point of application. In the biological wastewater treatment process, water-borne nutrients are assimilated into microbial biomass or biological sludge. The resulting sludge contains the nutrients assimilated in microbial biomass is further concentrated/dewatered. This leads to the generation of secondary wastewater stream containing high concentrations of reactive nitrogen, phosphorus, and organic compounds that need to be further treated.

Anaerobic digestion has been widely employed for treating concentrated waste streams such as manure and sludge generated in wastewater treatment processes, which contain high concentrations of nutrients. In some cases, nutrients in those waste streams have been recovered and returned to the agricultural fields by applying anaerobic digestion sludge that contains plant available forms of nutrients as a liquid fertilizer (Hospido et al. 2010; Sunaga et al. 2009). However, storing and transporting the nutrients contained in concentrated waste streams such as anaerobic digestion sludge are expensive due to their high water content coming with the target nutrients. It is important to concentrate the recovered nutrient to reduce the water content or ideally to separate the nutrients from the wastewater stream in order to lower the costs associated with the storage and transportation of the recovered nutrient products. The following section briefly introduces strategies of nitrogen and phosphorus recovery from concentrated waste streams.

4.1.2.1 Nitrogen Recovery

The most common strategy of nitrogen recovery from concentrated wastewaters is ammonia stripping by converting TAN (total ammonia nitrogen including ammonium and ammonia) contained in the waterbody into gaseous phase that can be stripped and recovered (Bonmatı and Flotats 2003, Saracco and Genon 1994, Sengupta et al. 2015). However, this strategy employs high solution pH that is above the pKa value of NH_3/NH_4^+, or 9.25, and high temperature to convert TAN in solution into gaseous phase. This ammonia-stripping operation is effective to remove TAN from wastewater body, but it consumes huge amount of energy and alkaline chemicals so that is costly and energy inefficient for treating most wastewaters.

As other nitrogen recovery strategies are concerned, ion exchange and adsorption-based approach have been examined to recover TAN from wastewaters. For example, the chemically modified zeolite is used to recover TAN from swine wastewater (Huang et al. 2014). Although some studies proposed to apply the nutrient-laden zeolite and other absorbents in agricultural fields as fertilizer, high cost of absorbents and usability of the recovered products as fertilizers are the major hurdles to be overcome for a large-scale implementation (Sengupta et al. 2015). The membrane-based separation operations have also been investigated to concentrate and separate TAN from wastewater. For instance, using the combination of reverse osmosis membrane separation and electrodialysis is effective in achieving highly concentrated nitrogen fertilizer that contains 13 g NH_3-N from each liter of liquid swine manure treated (Mondor et al. 2008). However, the capital investment and operational costs for the membrane wastewater treatment are too high to be practical for treating the majority of urban and animal wastewaters.

4.1.2.2 Phosphorus Recovery

Under high pH conditions (> 8.0), with the presence of equal amount of phosphate, ammonium, and magnesium, the soluble phosphorus will undergo spontaneous reaction to form magnesium ammonium phosphate (MAP) precipitates such as struvite (magnesium ammonium phosphate hexahydrate, $MgNH_4PO_4 \cdot 6H_2O$) (Le Corre et al. 2009), which is in plant-available form. Thus, chemical precipitation is an effective method to recover phosphorus from concentrated wastewaters instead of recovering the phosphorus as biosolids. Unlike other types of metal-phosphorus, i.e., iron-phosphorus or alum-phosphorus, MAP can be used as slow-release fertilizer with quality comparable to that of standard fertilizers such as diammonium phosphate and superphosphate (Rittmann et al. 2011). The feasibility of using MAP precipitates to recover phosphorus has been demonstrated with various types of wastewaters including swine wastewater (Song et al. 2007), digester residue from poultry manure (Yilmazel and Demirer 2011), and phosphorus mining wastewater (Hutnik et al. 2013). There are multiple commercially available configurations (Rittmann et al. 2011) of the MAP precipitation treatment. This treatment often requires the addition of magnesium amendment with amount equal to those of phosphate, ammonium, and magnesium in addition to maintaining the solution pH under alkaline condition to induce MAP precipitation. Thus, this is another costly and energy-intensive treatment to maintain high alkaline condition. Under similar chemical reaction conditions as for the MAP precipitation process, the magnesium (Mg) can be replaced by potassium (K), to form potassium ammonium phosphate precipitates. This is another application of the simultaneous N, P, and K recovery from wastewaters (Sengupta et al. 2015).

The aforementioned several chemical precipitation processes for nitrogen and phosphorus recovery from nutrient-concentrated wastewaters have been implemented in commercial sectors with promising performances. However, further developments are needed to improve the cost and energy efficiencies of those processes.

Recently, microbial electrochemical technologies (METs) that are based on various electrochemical reactions catalyzed by microorganisms attract growing interests as a novel approach to recover nutrients from concentrated wastewaters. The new MET treatment counts on bioelectrochemical reactions to achieve high pH conditions for facilitating the chemical precipitation process and selective transportation of the intended charged species. This will achieve savings in energy consumption and improved efficiency. Principles of METs and their recent developments on nutrient recovery will be introduced in latter sections.

Additionally, current infrastructures for collecting urban wastewater need to be improved for achieving more efficient nutrient recover by developing better nutrient recovery processes (Larsen et al. 2016). As a significant example, urine is a highly concentrated liquid source of N, P, and K. About 88% of human-excreted nitrogen and 67% of human-excreted phosphorus are present in urine for Swedish diet, whereas 70% of human-excreted nitrogen and 20–60% of human-excreted phosphorus are present in urine for Chinese diet (Mihelcic et al. 2011). Urine had been applied directly to agricultural field as fertilizer in conventional practice. In modern urban systems, however, urine mixed with feces and other wastewaters is treated as sewage. According to studies, 75% of nitrogen and 50% of phosphorus in domestic wastewaters are derived from human urine, but urine typically only accounts for less than 1% in volume of the wastewaters (Larsen and Gujer 1996, Mihelcic et al. 2011). Thus, separating the urine stream from other wastewater streams has been proposed as an effective method to use the separated urine as a premier source of the nutrient-containing stream by using the MET-based processes to recover nitrogen and phosphorus (Ledezma et al. 2015, 2017, Zamora et al. 2017).

4.2 Nutrient Recovery via Microbial Electrochemical Technologies (METs)

4.2.1 Principles and Applications of METs

Microbial electrochemical technologies (METs) are the hybrid technologies integrating microbial metabolism and electrochemical reactions for exploiting the capabilities of microbes to catalyze various electrochemical reactions to achieve simultaneous wastewater treatment and production of energy or chemicals (Logan and Rabaey 2012). For example, microbial fuel cells (MFCs) generate electrical power meanwhile oxidizing organic matter such as wasted biomass by using exoelectrogenic microbes that convert electrons stored in organic matter into the electrical current running through the external circuit. Exoelectrogenic microbes are living catalysts of electrochemical reactions in METs. The human body uses organic matter and oxygen as electron donor and acceptor, respectively, in the respiration process; organic matter is oxidized, whereas oxygen is reduced through electron transfer from organic matter to oxygen intracellularly to conserve energy. Exoelectrogenic microbes that perform anodic reaction of MFCs are capable of oxidizing organics and transferring the generated electrons out of the microbial cell to an insoluble electron acceptor. The energy is conserved by using insoluble metal oxides, which are abundant in subsurface environments, or artificial electrodes in engineered systems as an electron acceptor with such extracellular electron transfer (EET) reactions. METs exploit such EET reactions of electrogenic microbes to catalyze electrochemical reactions for various purposes. More information regarding those microbes and EET machineries can be found in recent review articles (Koch and Harnisch 2016, Kumar et al. 2017).

In an MFC, the bioanodic organic oxidation reaction where exoelectrogenic microbes oxidize organic compounds to generate electrons is coupled with the cathodic reduction reaction where terminal electron acceptors such as oxygen dissolved in the water are reduced by using anode-derived electrons (Fig. 4.2a). Electrons generated at the bioanode migrates to the cathode through an external circuit, and the resulting spontaneous electron flow from bioanode to cathode (electrical current) generates electrical power. The bioanode serves as an inlet of electron flow into the circuit, whereas the cathode serves as the outlet of electron flow from the circuit. By referencing a hydrogen fuel cell as an analogous system, hydrogen and organic compounds are the fuel (electron donor), and oxygen is the oxidizer (electron acceptor) in respective fuel cells, and microbes are the catalyst to accelerate the electrode reactions.

On the other hand, various chemicals can be produced by applying external electrical power into such microbial electrochemical system to drive otherwise nonspontaneous reactions. For example, microbial electrolysis cells (MECs) that generate hydrogen gas at the cathode through proton reduction employ a modified cathodic reaction of MFCs (Fig. 4.2b). If oxygen around the cathode in an MFC is purged, electron flow does not occur between bioanode and cathode because there

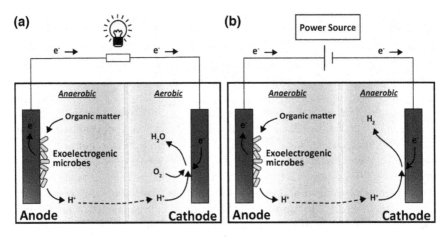

Fig. 4.2 Illustrations of representative setups of microbial electrochemical technologies. (**a**) Illustration of microbial fuel cell in which bioanodic organic oxidation is coupled with cathodic oxygen reduction resulting in electricity generation. (**b**) Illustration of microbial electrolysis cell in which anodic organic oxidation is coupled with cathodic proton reduction with external electrical power input resulting in hydrogen evolution

is no electron acceptor at the cathode to be spontaneously reduced under the condition. However, protons dissolved in the water around the cathode are reduced by accepting electrons to form hydrogen gas if the potential of an electrode is lower than −0.41 V vs. standard hydrogen electrode (SHE) under a relevant neutral pH condition (pH = 7). The electrode potential of the organic oxidizing bioanodes is around −0.28 V vs. SHE when acetate is oxidized. Thus, applying external electrical energy greater than −0.13 V between bioanode and cathode causes the cathode potential to drop below −0.41 V. Therefore, by forcing the cathode potential in this range with the externally applied electrical voltage induces spontaneous proton reduction, and an MEC enables electrons to flow from bioanode to cathode with proton reduction and hydrogen gas production (Logan et al. 2008). Practically, the voltage of −0.3 V or greater has been applied to secure the hydrogen production rate. The energy consumption by applying external voltage that is required to drive MEC reactions is typically much smaller than the energy retained in hydrogen gas produced in the reaction. In other words, MECs are analogous to water electrolysis that split water into oxygen gas at the anode and hydrogen gas at the cathode by applying external electrical energy. In this case, anodic water oxidation takes place at 0.81 V vs. SHE, whereas cathodic proton reduction takes place at −0.41 V vs. SHE under the neutral solution pH, and thus, the voltage of −1.22 V or greater is required to drive the water electrolysis reaction.

The MEC process requires less energy than the water electrolysis process due to favorable bioanode reaction, and this is a solid advantage over the latter for electrochemical hydrogen production. For converting biomass into hydrogen gas, laboratory-scale MEC tests demonstrated molar hydrogen yields of 8.55 mol H_2/mol hexose equivalent with glucose, 8.71 mol H_2/mol hexose equivalent with cellulose, and 3.65 mol H_2/mol acetate equivalent with acetic acid (Cheng and Logan 2007). These yields are significantly higher than the yields reported for fermentative hydrogen production of ~2 mol H_2/mol hexose (Hawkes et al. 2007) with the theoretical upper limit of 4 mol H_2/mol hexose. The MEC process can produce hydrogen from a wide variety of organic matter including various forms of volatile fatty acids that are hardly appropriate substrates to achieve effective fermentation. Additionally, almost all pure hydrogen gas can be recovered from a cathode chamber in which an ion-permeable membrane is used to separate the anode and cathode for excluding undesirable microbial reactions at the hydrogen-recovering cathode. As METs utilize both microbial electrochemical reactions and abiotic electrochemical reactions, an appropriate setup keeps microbial reactions and associated by-products away from contaminating the recovering products. This considerably reduces the subsequent efforts to purify the product, and this would be an advantage of METs over conventional bioprocesses that are often burdened by investment of cost and energy to separate the target products from other contaminants.

The ionic current caused by the migration of charged molecules in the electrolyte between anode and cathode generated by MFC and MEC reactions in the MET processes has also been exploited. The production of electrical current due to the flow of electrons through an external circuit in MFC/MEC is accompanied by the migration of charged species (ions) driven by electric field in the electrolyte between anode and cathode. As shown in Eqs. 4.4–4.8, the number of electron donated/accepted and electrical charge of ions consumed/generated and number of proton (or proton carriers) are balanced as the whole cell reaction, as well as the sum of bioanode and cathode are concerned.

Bioanodic organic oxidation in an MFC or an MEC (acetate as a model electron donor):

$$CH_3COO^- + 3H_2O \rightarrow CO_2 + HCO_3^- + 8H^+ + 8e^- \qquad (4.4)$$

Cathodic oxygen reduction in MFC:

$$2O_2 + 8e^- + 8H^+ \rightarrow 4H_2O \qquad (under\,pH < 7) \qquad (4.5)$$

$$2O_2 + 8e^- + 4H_2O \rightarrow 8OH^- \qquad (under\,pH > 7) \qquad (4.6)$$

Cathodic proton reduction (hydrogen evolution) in MEC:

$$8H^+ + 8e^- \rightarrow 4H_2 \qquad (under\,pH < 7) \qquad (4.7)$$

$$8H_2O + 8e^- \rightarrow 4H_2 + 8OH^- \qquad (under\,pH > 7) \qquad (4.8)$$

For example, when the MFC is fueled by acetate with oxygen reducing cathode under acidic condition, the bioanode reaction donates eight electrons (per an acetate) to the circuit and releases eight protons into the electrolyte. Then, at the cathode, the eight electrons from the circuit and the eight protons from the electrolyte are consumed in the oxygen reduction reaction. As the electrical current flowing through the circuit secures electron balance, migration of charged ions in the electrolyte is driven by the electric field and diffusion between anode and cathode thus serving a role in charge/proton balancing. When electrons move from anode to cathode through the external circuit, positively charged ions (cations) migrate from anode to cathode through the electrolyte.

With an appropriate arrangement of ion-exchange membranes, this current-driven ionic migration can be used for selectively transporting or separating charged molecules contained in the electrolyte. Ion-exchange membranes are a type of semipermeable membrane that allow certain types of ions to pass through while blocking other types of ions as well as uncharged neutral molecules. If the anode chamber and the cathode chamber in an MFC/MEC are separated by a cation-exchange membrane (CEM), only cations are selectively transported through the membrane while anions are rejected. Cations such as H^+, Na^+, Ca^{2+}, or NH_4^+ in the solution will migrate from the anode chamber to the cathode chamber so that they are retained and accumulated in the cathode chamber or consumed by the cathodic reaction. On the other hand, when an anion-exchange membrane (AEM) separates the anode and cathode chambers, cation migration from anode chamber to cathode chamber is rejected by the membrane; instead, anions are allowed to migrate from cathode chamber to anode chamber to maintain charge neutrality in both chambers. If the anode and cathode chambers are separated by a pair of CEM and AEM to create a third chamber arranged in the order of anode chamber, CEM, third chamber, AEM, and cathode chamber, the current through the circuit causes cations to transport from anode chamber into the middle chamber and anions to transport from cathode chamber into the middle chamber. As the result, both anions and cations are retained and concentrated into the middle chamber as MFC/MEC reactions continues (Fig. 4.3a). Such a setup is defined as bio-electroconcentration system that has been studied for concentrating valuable charged species such as NH_4^+, K^+, and phosphate ion (Ledezma et al. 2017). In contrast, the arrangement of anode chamber, AEM, third chamber, CEM, and cathode chamber makes a different system. Namely, the concentration of ions in the middle chamber is decreased continuously because anions move from the middle chamber into the anode chamber, whereas cations move from middle to the cathode chamber (Fig. 4.3b). This setup has been researched to remove the charged species from a solution placed in the middle chamber such as in the desalination of sea/brackish water to produce fresh water. The solution to be desalinated is loaded in the middle chamber to get rid of anions and cations. Studies of this process, which is known as the "microbial desalination cells" (Cao et al. 2009), show that it is a potential novel energy-offsetting desalination process with simultaneous wastewater treatment (Kim and Logan 2013). These MET-based separation processes to concentrate or reduce concentrations of specific charged species in a solution body based on semipermeable membranes and electric field-driven migration of charged species are classified as a branch of electrodialysis process used in water purification and food/drink processing.

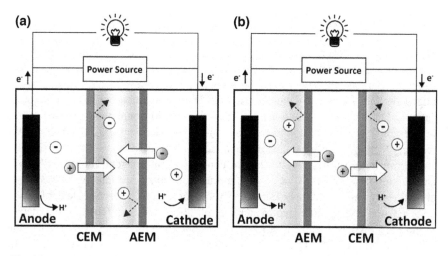

Fig. 4.3 Illustrations of ionic species separation processes in an MFC or an MEC. (**a**) Arrangement of Anode-CEM-AEM-cathode to concentrate ions in the middle chamber. (**b**) Arrangement of Anode-AEM-CEM-cathode to de-concentrate ions in the middle chamber. CEM and AEM represents cation exchange membrane and anion exchange membrane, respectively

This section introduced the principles and applications of several aspects of METs. The MFC/MEC processes for nitrogen and phosphorus recovery using bio-electrochemical reactions and specific membrane to perform selective ion transport and separation as introduced in above paragraphs will be reviewed in the next section of this chapter.

4.2.2 Microbial Electrochemical Technologies (METs) for Nutrient Recovery

METs, which were recently implemented in the commercial sector, are emerging technologies to degrade wasted biomass while producing energy and/or chemicals. Focuses of the METs being developed as a wastewater treatment technology are primarily on decomposition of organic matter in wastewaters, but some recent MET designs address nutrient removal/recovery so that this process will provide a novel nutrient management strategy for treating wastewater streams. During the last decade, many studies have been conducted on integrating nutrient removal processes with METs. For example, the nitrogen removal process applying the integrated nitrification and denitrification processes were incorporated in METs to result in a treatment with low energy requirements (Kelly and He 2014). More importantly, other studies applying METs process to recover nitrogen and phosphorus from waste streams have been demonstrated to be a potential alternative energy-efficient nutrient recovery process. This section reviews recent advances and future perspectives of MET-based nutrient recovery processes.

4.2.2.1   Nitrogen Recovery

The process to recover nitrogen in METs mainly targets TAN including ammonium and ammonia. Separation and concentration of the recovered ammonia from the solution in METs will be focused in this section. It is accomplished by using cation migration that is driven by the electric field and the alkaline condition around the cathode due to cathodic reactions. Incorporation of additional membrane processes in last few years greatly enhances the feasibility to scale up the MET-based TAN recovery technology for a full-scale application.

Recovering TAN from waste streams containing low concentrations of TAN is a challenging task. Even for wastes containing high concentrations of TAN such as animal wastewaters or digester sludge, TAN separation for a large-scale application is costly and energy intensive (Maurer et al. 2006). To address the issue, cation migration in METs has been utilized to concentrate ammonium. The cations that migrate in MFCs/MECs to offset the charge imbalance caused by the generation of protons at anode and consumption of protons at cathode are typically not protons but other cations such as sodium ions and ammonium ions which have much higher concentrations in the wastewaters. A study reported that ammonium acts as the proton shuttle, and its migration from anode to cathode through the CEM with separated anode chamber and cathode chamber accounts for ~90% of the total ionic flux in an MET (Cord-Ruwisch et al. 2011). By using this ammonium ion migration through CEM, TAN can be separated from the anode chamber and concentrated in the cathode chamber (Kuntke et al. 2011). According to a study, this ammonium migration in an MEC results in about ten times concentrated TAN solution of over 300 mg N/L at the cathode (Villano et al. 2013). Production of concentrated TAN solution by intrinsic cation migration and CEM in an MET alleviates the cost and energy demand in TAN recovery. The concentrated TAN solution thus produced in the cathode chamber can be used as a liquid nitrogen fertilizer. However, further research is needed to evaluate the feasibility and issues of using these solutions as a fertilizer.

Several different methods have been reported in literature to further refine the concentrated TAN solution collected in the cathode chamber. The first approach is stripping the ammonia with gas sparged through the TAN concentrate. But a major issue of the method is its high energy demand for sparging gas as well as maintaining high temperature and high pH conditions to favor the partition of ammonia into gaseous phase (Maurer et al. 2006). In this MET process, however, the required high pH condition is achieved not by adding external alkali as in the conventional ammonia stripping but by using the intrinsic alkaline production at the cathodes without adding additional chemicals. The cathodic reactions such as oxygen reduction in MFCs (Eqs. 4.5 and 4.6) and hydrogen evolution in MECs (Eqs. 4.7 and 4.8) raise the pH of the solution particularly around the cathode above the pKa of NH_3/NH_4^+, or 9.25, which is needed to convert the soluble ammonium (NH_4^+) into ammonia gas (NH_3).

Using the method, ammonia is recovered and purified from urine in an MFC under high catholyte pH followed by aeration leading to subsequent adsorption in acid solution (Kuntke et al. 2012). In an MEC, the catholyte pH can be increased above 12 due to high current generation; ammonia was stripped with simultaneous production of hydrogen gas with recovery efficiency of ~80% using a real ammonium-rich wastewater sample (Wu and Modin 2013). The results obtained in a more recent work show that ammonium migration and stripping are further enhanced by the presence of NaCl in catholyte to achieve 94.3% recovery of ammonium from pig slurry containing 523 mg $N-NH_4^+/L$ (Sotres et al. 2015).

In another approach, TAN recovery was carried out by absorption in sulfuric acid solution through a gas-permeable hydrophobic membrane known as TMCS (trans-membrane chemisorption) (Kuntke et al. 2016). This is the ammonia stripping not by sparging gas but using a gas-permeable hydrophobic membrane. The TMCS could lower the energy requirement for TAN recovery because it does not need a gas sparging with high flow rate of several L of gas per L of wastewater to strip ammonia out of the solution (Kuntke et al. 2016). The driving force of TMCS is the ammonia concentration gradient across the gas-permeable membrane that requires a high pH (>8.5) in the feed solution and lower pH (<7) in the solution on the other side of the membrane (Kuntke et al. 2018). Such pH conditions can be easily maintained by using catholyte as the feed solution and an appropriate acid solution on the product side. The consistent performance of a recent work using a scaled-up MEC with TMCS module demonstrates the potential of this treatment method as an energy-efficient TAN recovery from urine (Zamora et al. 2017).

Furthermore, the three-chamber MEC with the arrangement of anode, CEM, AEM, and cathode, which is described as the bio-electroconcentration system in previous sections, was examined for recovering TAN from urine (Ledezma et al. 2017). In this system, ammonium ions are transported from the anolyte to the middle chamber. Simultaneously, bicarbonate ions were transported to the middle chamber from the catholyte, which is anode effluent passed through a subsequent treatment removing organic matters. As a result, ammonium bicarbonate salt precipitate is recovered in the concentrated solution at the middle chamber (Ledezma et al. 2017). Although this concept of recovering TAN from organic wastes has been proved feasible using laboratory-scale equipment, further research is needed to evaluate the potential of a full-scale implementation of this novel process for TAN recovery.

The above discussions reveal that the METs system has the potential to concentrate TAN in wastewaters in an energy-efficient manner by using electrical current to drive the ammonia migration. Also, the alkaline condition derived by cathode reactions requires less energy consumption and alkali chemical demand for recovering ammonia from the concentrated solution by using various strategies such as ammonia stripping, TMCS, and precipitation of ammonia salts such as struvite and ammonium bicarbonate. Collectively, the MET-based TAN recovery processes show potentials as an energy-efficient alternative nitrogen recovery technology from various waste streams. However, most of the literature results are obtained using laboratory-scale experiments; further research is needed to enhance the mechanistic understanding of these processes as the bases for developing on-site treatment systems.

4.2.2.2 Phosphorus Recovery

The recovery of phosphorus in METs has been reported in the form of precipitation of magnesium ammonium phosphate (MAP) crystals such as struvite, $MgNH_4PO_4 \cdot 6H_2O$, on the cathode surface of MFCs/MECs. The mechanism of MAP crystal precipitation on cathode surface suggested in those literature is alkali pH conditions produced by cathode reactions and the supply of ammonium ion and magnesium ion moving to the cathode caused by cation migration. Using a single-chamber MEC, the struvite precipitation achieves the maximum 40% of phosphate removal from a synthetic ammonium-phosphorus-rich electrolyte at a precipitation rate of 0.3–0.9 $g/(m^2 h)$ (Cusick and Logan 2012). In a subsequent study, a two-chamber MEC with fluidized bed cathode chamber was used to evaluate struvite precipitation and simultaneous hydrogen production from digester effluent (Cusick et al. 2014). The results indicate that struvite precipitation results in 70–85% removal of soluble phosphorus, and the scouring provided by the fluidized particles prevents the scale accumulation on the cathode. Additionally, the results also prove that significant savings in energy consumption can be achieved by using the tested MEC process as compared with other pH adjustment approaches. This shows the potential of MET-based MAP crystal precipitation as an energy-efficient process to recover phosphorus from organic wastes (Cusick et al. 2014).

A single-chamber air-cathode MFC was tested to remove organic matter and phosphorus from swine wastewater in addition to generating electricity (Ichihashi and Hirooka 2012). About 70–80% of phosphorus is removed from the influent, and the amount of phosphorus precipitated (mainly struvite) at the cathode was equivalent to 5–27%. The subsequent study carried out by the same research group reported that amendments of ammonium, phosphate, and magnesium influence formation of the precipitate that covers the cathode surface and decreases the cathode performance (Hirooka and Ichihashi 2013). Recently, phosphorus remobilization of iron phosphate in digested sewage sludge and subsequent precipitation as struvite precipitates was performed by using a 3-L MET reactor operated as MFC and also as MEC (Happe et al. 2016).

As reviewed above, there are only a limited number of studies addressing phosphorus recovery with METs reported in literature. Although results of those studies demonstrate that the process has some advantages such as requiring less energy inputs, the reported findings also reveal that some limitations and potentials of the MET-based phosphorus recovery are still largely unknown. Hence, significant research focus needs to be directed toward issues including the potential recovery of other species of phosphorus precipitate such as calcium phosphate precipitation reported in electrolysis process (Lei et al. 2017), other non-MAP species, negative effects of long-term MAP precipitation that potentially blocks the catalytic sites on the cathode resulting in decreased MET performances, strategy to recover the precipitate, and quality of recovered precipitates to be used as fertilizers. More importantly, the development of promising strategies for recovering precipitates and regeneration/replacement of the degenerated cathode electrode would be crucial issues because the precipitate is expected to form on cathode surface. Development of expendable cathode electrode that can be applied directly to the agricultural field would be a potential alternative option. Another alternative is the use of cathodic electrode made of activated carbon or biochar that can be completely removed from the process with phosphorus precipitates and applied the mixture as solid fertilizer to a field.

Box 4.1 Electrode Versus Nitrate; to Develop Integrated Nitrogen Management Strategy in Microbial Electrochemical Technologies (METs)

Microbial anode reduction in which electron transfers to the anode electrode by exoelectrogenic microbes is a key factor to influence the MET performance. In MET anode systems, exoelectrogens use anode electrode as the electron acceptor where electrons that pass through the intracellular electron transport chain caused by respiration are discharged. However, many exoelectrogens are known to respire with different chemical species such as oxygen or nitrate other than the electrode as an electron acceptor. Thus, purging other competing electron acceptors from the system is of great importance to warrant that the anode electrode is the sole electron acceptor for exoelectrogens. However, recent MET development with integrated nitrogen removal/recovery processes bring about an opportunity to introduce oxidized nitrogen compounds such as nitrate on exoelectrogens in the anode system. Some exoelectrogens such as *Geobacter* spp., *Shewanella* spp., and *Geoalkalibacter* spp. are capable of reducing nitrate, which is the most energetically favorable electron acceptor in anoxic conditions, and often control anoxic microbial reactions. Nitrate could act as a competitive electron acceptor to the anode, but its negative effects on exoelectrogens and MET performances have been largely overlooked. Therefore, understanding their facultative metabolisms, as well as the anode reduction versus potential nitrate reduction, is important to develop reliable MET processes as the strategy to achieve nitrogen removal/recovery.

A study focused on understanding the critical conditions controlling the facultative anode versus nitrate metabolisms for an exoelectrogenic biofilm grown on anode electrodes in an MET reactor by using *Geobacter metallireducens*, a model exoelectrogen capable of nitrate reduction (Kashima and Regan 2015). The findings are presented in the following sections. Based on Nernst equation-derived thermodynamic energy gains for anode and nitrate reduction, factors that are hypothesized to control these facultative metabolisms include nitrate concentration in bulk solution, biofilm thickness, and also the thermodynamic availability of the anode as determined by anode potential. To test this hypothesis, *G. metallireducens* biofilms of various biofilm thickness are acclimated on graphite block anode electrodes in two-chamber MET reactors potentiostatically controlled at different levels of

(continued)

Box 4.1 (continued)

anode potential. The systems were subjected to nitrate spikes over a range of concentrations to measure the anode and nitrate reduction rates with respect to nitrate concentration.

The results show that the anode-reducing *G. metallireducens* biofilm preferentially reduces nitrate. Under all tested anode potentials ($-150 \sim +900$ mV vs. SHE), the metabolism switches from anode reduction to nitrate reduction in response to nitrate spikes. This illustrates that nitrate has negative effects on anode performance by acting as a competitive electron acceptor to the anode (Fig. 4.4a).

Additionally, the nitrate availability but not the anode availability controls whether anode or nitrate is used as the electron acceptor in the facultative metabolism. Specifically, the critical nitrate concentration that triggers a striking decrease in anode biofilm-generated current increases as a function of biofilm thickness, whereas no effect was observed among different anode potentials. This suggests that the nitrate effect as a competing electron acceptor to the anode is controlled by nitrate concentration in the bulk solution and its diffusion into the anode-reducing biofilm (Fig. 4.4b).

Concerning the MET performance, a nitrate concentration greater than the critical level leads to striking malfunction of the anodic performance. The production of current for anodes exposed to nitrate with concentrations above the critical level significantly decreases to the background level until nitrate and nitrite are consumed. The accumulation of nitrite, which is an intermediate product through dissimilatory nitrate reduction reactions, is toxic to microbial cells thus decreasing the entire biological activity of the biofilm. Furthermore, nitrate in the anode chamber lowers the coulombic recovery and induces suspended biomass growth. Increasing suspended biomass would be problematic when the wastewater treatment efficiency is concerned because these suspended biomass needs to be further removed with additional treatment processes (Fig. 4.4c) for the treated effluent to meet stringent quality limitations. These results provide important fundamental information about facultative nitrate reduction versus anode reduction in METs for developing a stable nitrogen removal/recovery strategy in METs.

(continued)

Box 4.1 (continued)

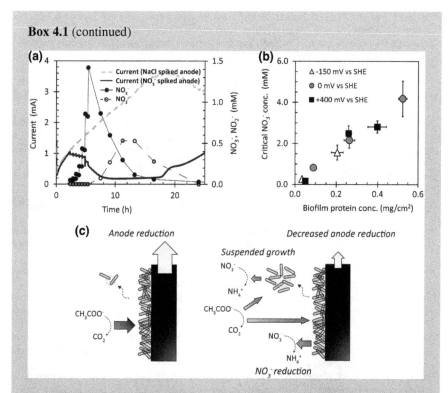

Fig. 4.4 Facultative nitrate reduction by anode biofilms. (**a**) Current, nitrate, and nitrite profiles in a gradual nitrate spike test of *Geobacter metallireducens* biofilms acclimatized on anodes poised at 0 mV vs. SHE. (**b**) Critical nitrate conc. (nitrate conc. in bulk liquid that triggers current decrease) of anode biofilms with different biofilm thicknesses at different anode potentials. (**c**) Schematic graphical presentation of flows of acetate-derived electrons in bioanode systems with and without the influence of nitrate

References

Bennett EM, Carpenter SR, Caraco NF (2001) Human impact on erodable phosphorus and eutrophication: A global perspective increasing accumulation of phosphorus in soil threatens rivers, lakes, and coastal oceans with eutrophication. Bioscience 51:227–234. https://doi.org/10.1641/0006-3568(2001)051[0227:HIOEPA]2.0.CO;2

Bonmatı A, Flotats X (2003) Air stripping of ammonia from pig slurry: characterisation and feasibility as a pre-or post-treatment to mesophilic anaerobic digestion. Waste Manag 23:261–272

Canfield DE, Glazer AN, Falkowski PG (2010) The evolution and future of Earth's nitrogen cycle. Science 330:192–196

Cao X, Huang X, Liang P, Xiao K, Zhou Y, Zhang X, Logan BE (2009) A new method for water desalination using microbial desalination cells. Environ Sci Technol 43:7148–7152

Cheng S, Logan BE (2007) Sustainable and efficient biohydrogen production via electrohydrogenesis. Proc Natl Acad Sci 104:18871–18,873

Clauwaert P et al (2007) Biological denitrification in microbial fuel cells. Environ Sci Technol 41:3354–3360

Cordell D, Drangert J-O, White S (2009) The story of phosphorus: global food security and food for thought. Glob Environ Chang 19:292–305

Cord-Ruwisch R, Law Y, Cheng KY (2011) Ammonium as a sustainable proton shuttle in bioelectrochemical systems. Bioresour Technol 102:9691–9696

Cusick RD, Logan BE (2012) Phosphate recovery as struvite within a single chamber microbial electrolysis cell. Bioresour Technol 107:110–115

Cusick RD, Ullery ML, Dempsey BA, Logan BE (2014) Electrochemical struvite precipitation from digestate with a fluidized bed cathode microbial electrolysis cell. Water Res 54:297–306

De-Bashan LE, Bashan Y (2004) Recent advances in removing phosphorus from wastewater and its future use as fertilizer (1997–2003). Water Res 38:4222–4246

Fixen PE, West FB (2002) Nitrogen fertilizers: meeting contemporary challenges. AMBIO J Hum Environ 31:169–176

Happe M et al (2016) Scale-up of phosphate remobilization from sewage sludge in a microbial fuel cell. Bioresour Technol 200:435–443

Hawkes FR, Hussy I, Kyazze G, Dinsdale R, Hawkes DL (2007) Continuous dark fermentative hydrogen production by mesophilic microflora: principles and progress. Int J Hydrog Energy 32:172–184

Hirooka K, Ichihashi O (2013) Phosphorus recovery from artificial wastewater by microbial fuel cell and its effect on power generation. Bioresour Technol 137:368–375

Hospido A, Carballa M, Moreira M, Omil F, Lema JM, Feijoo G (2010) Environmental assessment of anaerobically digested sludge reuse in agriculture: potential impacts of emerging micropollutants. Water Res 44:3225–3233

Huang H, Xiao D, Pang R, Han C, Ding L (2014) Simultaneous removal of nutrients from simulated swine wastewater by adsorption of modified zeolite combined with struvite crystallization. Chem Eng J 256:431–438

Hutnik N, Kozik A, Mazienczuk A, Piotrowski K, Wierzbowska B, Matynia A (2013) Phosphates (V) recovery from phosphorus mineral fertilizers industry wastewater by continuous struvite reaction crystallization process. Water Res 47:3635–3643

Ichihashi O, Hirooka K (2012) Removal and recovery of phosphorus as struvite from swine wastewater using microbial fuel cell. Bioresour Technol 114:303–307

Jetten MS et al (1998) The anaerobic oxidation of ammonium. FEMS Microbiol Rev 22:421–437

Kashima H, Regan JM (2015) Facultative nitrate reduction by electrode-respiring Geobacter metallireducens biofilms as a competitive reaction to electrode reduction in a bioelectrochemical system. Environ Sci Technol 49:3195–3202

Kelly PT, He Z (2014) Nutrients removal and recovery in bioelectrochemical systems: a review. Bioresour Technol 153:351–360

Kim Y, Logan BE (2013) Microbial desalination cells for energy production and desalination. Desalination 308:122–130

Koch C, Harnisch F (2016) Is there a specific ecological niche for electroactive microorganisms? Chem Electro Chem 3:1282–1295

Kumar A et al (2017) The ins and outs of microorganism–electrode electron transfer reactions. Nat Rev Chem 1:0024

Kuntke P, Geleji M, Bruning H, Zeeman G, Hamelers H, Buisman C (2011) Effects of ammonium concentration and charge exchange on ammonium recovery from high strength wastewater using a microbial fuel cell. Bioresour Technol 102:4376–4382

Kuntke P et al (2012) Ammonium recovery and energy production from urine by a microbial fuel cell. Water Res 46:2627–2636

Kuntke P, Zamora P, Saakes M, Buisman C, Hamelers H (2016) Gas-permeable hydrophobic tubular membranes for ammonia recovery in bio-electrochemical systems. Environ Sci Technol Lett 2:261–265

Kuntke P et al (2018) (Bio) electrochemical ammonia recovery: progress and perspectives. Appl Microbiol Biotechnol 102:3865–3878

Lackner S, Gilbert EM, Vlaeminck SE, Joss A, Horn H, van Loosdrecht MC (2014) Full-scale partial nitritation/anammox experiences–an application survey. Water Res 55:292–303

Larsen TA, Gujer W (1996) Separate management of anthropogenic nutrient solutions (human urine). Water Sci Technol 34:87–94

Larsen TA, Hoffmann S, Lüthi C, Truffer B, Maurer M (2016) Emerging solutions to the water challenges of an urbanizing world. Science 352:928–933

Le Corre KS, Valsami-Jones E, Hobbs P, Parsons SA (2009) Phosphorus recovery from wastewater by struvite crystallization: A review. Crit Rev Environ Sci Technol 39:433–477

Ledezma P, Kuntke P, Buisman CJ, Keller J, Freguia S (2015) Source-separated urine opens golden opportunities for microbial electrochemical technologies. Trends Biotechnol 33:214–220

Ledezma P, Jermakka J, Keller J, Freguia S (2017) Recovering nitrogen as a solid without chemical dosing: bio-electroconcentration for recovery of nutrients from urine. Environ Sci Technol Lett 4:119–124

Lei Y, Song B, van der Weijden RD, Saakes M, Buisman CJ (2017) Electrochemical induced calcium phosphate precipitation: importance of local pH. Environ Sci Technol 51:11156–11,164

Liu Y, Villalba G, Ayres RU, Schroder H (2008) Global phosphorus flows and environmental impacts from a consumption perspective. J Ind Ecol 12:229–247

Logan BE, Rabaey K (2012) Conversion of wastes into bioelectricity and chemicals by using microbial electrochemical technologies. Science 337:686–690

Logan BE, Call D, Cheng S, Hamelers HV, Sleutels TH, Jeremiasse AW, Rozendal RA (2008) Microbial electrolysis cells for high yield hydrogen gas production from organic matter. Environ Sci Technol 42:8630–8640

Maurer M, Pronk W, Larsen T (2006) Treatment processes for source-separated urine. Water Res 40:3151–3166

McCarty PL, Bae J, Kim J (2011) Domestic wastewater treatment as a net energy producer–can this be achieved? Publications, ACS

Mihelcic JR, Fry LM, Shaw R (2011) Global potential of phosphorus recovery from human urine and feces. Chemosphere 84:832–839

Mondor M, Masse L, Ippersiel D, Lamarche F, Masse D (2008) Use of electrodialysis and reverse osmosis for the recovery and concentration of ammonia from swine manure. Bioresour Technol 99:7363–7368

Mosier A, Duxbury J, Freney J, Heinemeyer O, Minami K (1998) Assessing and mitigating N_2O emissions from agricultural soils. Clim Chang 40:7–38

Nancharaiah Y, Mohan SV, Lens P (2016) Recent advances in nutrient removal and recovery in biological and bioelectrochemical systems. Bioresour Technol 215:173–185

Ravishankara A, Daniel JS, Portmann RW (2009) Nitrous oxide (N_2O): the dominant ozone-depleting substance emitted in the 21st century. Science 326:123–125

Rittmann BE, Mayer B, Westerhoff P, Edwards M (2011) Capturing the lost phosphorus. Chemosphere 84:846–853

Saracco G, Genon G (1994) High temperature ammonia stripping and recovery from process liquid wastes. J Hazard Mater 37:191–206

Sengupta S, Nawaz T, Beaudry J (2015) Nitrogen and phosphorus recovery from wastewater. Curr Pollut Rep 1:155–166

Smil V (2002) Nitrogen and food production: proteins for human diets. AMBIO J Hum Environ 31:126–131

Song Y, Yuan P, Zheng B, Peng J, Yuan F, Gao Y (2007) Nutrients removal and recovery by crystallization of magnesium ammonium phosphate from synthetic swine wastewater. Chemosphere 69:319–324

Sotres A, Cerrillo M, Viñas M, Bonmatí A (2015) Nitrogen recovery from pig slurry in a two-chambered bioelectrochemical system. Bioresour Technol 194:373–382

Strous M et al (1999) Missing lithotroph identified as new planctomycete. Nature 400:446–449

Sunaga K et al (2009) Impacts of heavy application of anaerobically digested slurry to whole crop rice cultivation in paddy environment on water, air and soil qualities. Jpn J Soil Sci Plant Nutr 80:596–605

Villano M, Scardala S, Aulenta F, Majone M (2013) Carbon and nitrogen removal and enhanced methane production in a microbial electrolysis cell. Bioresour Technol 130:366–371

Wu X, Modin O (2013) Ammonium recovery from reject water combined with hydrogen production in a bioelectrochemical reactor. Bioresour Technol 146:530–536

Yilmazel Y, Demirer G (2011) Removal and recovery of nutrients as struvite from anaerobic digestion residues of poultry manure. Environ Technol 32:783–794

Zamora P et al (2017) Ammonia recovery from urine in a scaled-up microbial electrolysis cell. J Power Sources 356:491–499

Chapter 5
Energy Production from Wasted Biomass

Miftahul Choiron, Seishu Tojo, and Megumi Ueda

Abstract The production of wastes in various forms is the consequence of human activities, and the waste-related problems are becoming more serious with increasing human population and activities. The degradable organic waste originated from agricultural products and foods contains recoverable energy existing in many forms of components. There are several waste-to-energy technologies including bio-gasification/anaerobic digestion available for recovering the energy.

Recent energy research emphasizes issues on the depletion of fossil fuel and adverse environmental impacts. The extensive use of fossil fuel causes the emission of a large quantity of waste gas known as the greenhouse gas. Using renewable energy is expected to not only provide energy but also solve other problems such as waste discharge, gas emission, and global warming, among others. Hydrogen gas is universally recognized as an environmentally safe and renewable energy. The combustion of hydrogen gas produces only water, thus making it an ideal alternative energy to fossil fuels. Hot compressed water (HCW) is the condition of liquid water when it is subject to elevated temperature and pressure. Many agricultural and food industrial wastes containing cellulose or hemicellulose are degraded with HCW treatment at high temperature and pressure into soluble oligomers. An attempt of biohydrogen production from biomass by using HCW method as pretreatment is explained.

Microbial fuel cell (MFC) generates electricity by extracting the electronic current directly from organic matter biologically. However, the biological process to decompose organic matter and produce electricity is time-consuming. The integrated MFC system consists of producing hydrogen gas as a combustible fuel by

M. Choiron
Department of Agro-industrial Technology, University of Jember, Jember, Indonesia
e-mail: m.choiron@unej.ac.id

S. Tojo (✉)
Institute of Agriculture, Tokyo University of Agriculture and Technology, Fuchu, Tokyo, Japan
e-mail: tojo@cc.tuat.ac.jp

M. Ueda
United Graduate School of Agricultural Science, Tokyo University of Agriculture and Technology, Fuchu, Tokyo, Japan

© Springer Nature Singapore Pte Ltd. 2020   91
S. Tojo (ed.), *Recycle Based Organic Agriculture in a City*,
https://doi.org/10.1007/978-981-32-9872-9_5

hydrogen fermentation in the first stage and generating electricity by MFC in the secondary stage. The biological hydrogen fermentation in the first stage also produces organic acids that can be used as the substrate for generating electricity in the second stage. The MFC system is an adaptable technology to meet the power demand needed for operating the agricultural systems and facilities in urban regions.

5.1 Energy Potential of Wasted Biomass

5.1.1 Wasted Biomass in City

The production of wastes in various forms is the consequence of human activities, and the waste-related problems are becoming more serious with increasing human population and activities. Waste recycling is one of numerous methods to address environmental issues. Inappropriate waste management was regarded a problem in the past, but it can also be considered a challenge in the future to produce renewable energy and materials from the waste. "Biomass," which is defined as any organic matter available on a renewable basis, comes from plants, animal wastes, microalgae, and many others. There are three sources of biomass: (1) residues and waste including municipal, agricultural, forestry, and industrial wastes, (2) forestry including natural and seminatural forests and plantations, and (3) crops and fast-growing grasses (IEA and FAO 2017). Energy contained in crops and fast-growing grasses can be recovered as "bioenergy," but a practical application may interfere with adequate supply of food and feed.

In urban regions, most of the waste is from municipal, agricultural, and industrial sources. Municipal wastes in a combination of solid and semisolid forms come from household, traditional/modern market, office, street, hospital, commercial area, and others. Among these wastes, the degradable organic waste originates from the agri-

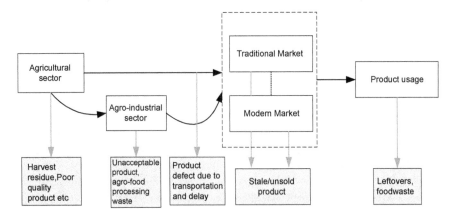

Fig. 5.1 Organic wastes generation from agricultural activities in urban regions

cultural sector including livestock and seafood productions (Fig. 5.1). Additionally, the organic waste may be produced when agricultural and agro-industrial products are transported to market in the urban regions.

The municipal organic waste in different nations may have various characteristics depending on whether a nation is a developing or developed country. Cities from developing country produce more putrescible waste than cities from developed country (UNEP 2005). In general, the degradable municipal organic waste that originates from agricultural products such as livestock and/or seafood contains recoverable energy existing in many forms of components. There are several *waste-to-energy* technologies including bio-gasification/anaerobic digestion available for recovering the energy.

Researches on municipal solid waste management focus on the collection and transportation of the waste in addition to waste treatment and processing. In Japan, solid wastes are collected by using small low-pollution trucks and transported to transfer stations where they are compressed and then loaded into containers. The container is then transported in a large truck to a disposal site or incineration plant. This practice assists in reducing both operating cost and CO_2 emission (JESC 2012). An alternative strategy to improve the efficiency of biomass collecting and transporting is to optimize the organic waste vehicle route (Das and Bhattacharyya 2015; Ramos et al. 2018).

5.1.2   Waste Management in Developing Country: Case Study in Jember City, Indonesia

Municipal and district governments are responsible for managing municipal solid wastes. Most municipal governments in Indonesia implement landfill to manage municipal waste. Jember is a city located in East Java province. Similar to other cities in Indonesia, Jember uses landfill as the primary method to manage the waste

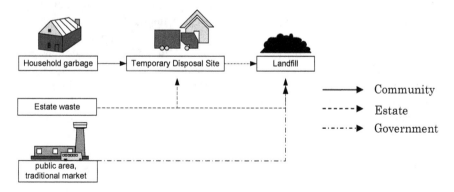

Fig. 5.2 Waste collection and transportation practiced in Indonesia

generated by household, estate, public area or market, and the waste collected and transported by different organizations (Fig. 5.2).

The household waste is generally not separated so that the garbage bag left in front of residential households contains mixed organic and inorganic components. These bags are collected and transported to a temporary disposal site in garbage dump trucks daily. The service is provided and paid for by neighborhood and community organizations periodically. The waste is then transported from the temporary disposal site in trucks to a landfill site where organic waste is separated and disposed of in a controlled landfill; degradable biomass or organic material is spread on the lots at about 60–100 cm thick. It is compacted and then covered with a layer of soil of about 40 cm in thickness (Khoiron 2014). On average, microbial anaerobic activities produce and release methane gas, which can be collected and distributed to households around the landfill site through biogas channel piping.

The municipal waste management practice in developed countries are different from those in developing countries. Using Fuchu city in Tokyo, Japan, as an example, landfill is infrequently practiced as the municipal solid waste management method because of limited available land space. Hence, the separated waste is collected from each household following a predetermined schedule. The waste bag color is used as the code to distinguish combustible wastes (green), plastic waste (pink), and noncombustible wastes (orange). Other noncombustible wastes, such as can, recyclable goods, glass, or PET bottle, are also picked up separately. Some other cities in Japan practice advanced recycling technology and waste treatment for managing wasted biomass.

Box 5.1 Biogas Production in Miura Biomass Center, Japan

Japan has a leading position on green energy generation. Efficient technologies on waste transport, incineration technology, PET bottle and home appliance recycling technology, as well as waste landfill and medical waste disposal, and biomass utilization have been implemented as the strategy for achieving sustainable environment (JESC 2012). Miura Biomass Center, which is located in Miura city, Kanagawa Prefecture, Japan, is an integrated facility for biogas production and wastewater treatment.

Biogas facilities obtain raw materials at 5.5 t/d for agricultural crop harvest residue, 0.5 t/d for aquatic residues, and 55 kL/d for human waste and sewage sludge. The fermentation process is operated in mesophilic temperature with hydraulic retention time of 14 days. Approximately 1000 m^3/day biogas is produced with about 60% methane concentration. Biogas is combusted to generate 25 kW electricity and 38 kW thermal energy. The digested slurry is processed into 1.4 t/d compost fertilizer through composting for 14 days that can be applied on farmlands (Fig. 5.3). The operational sustainability is supported by collaboration among farmers, companies, and government. Farmers and companies sent their organic waste (crop residues, sewage sludge etc.) to a biogas plant, and farmers can purchase fertilizer from the plant.

(continued)

Box 5.1 (continued)

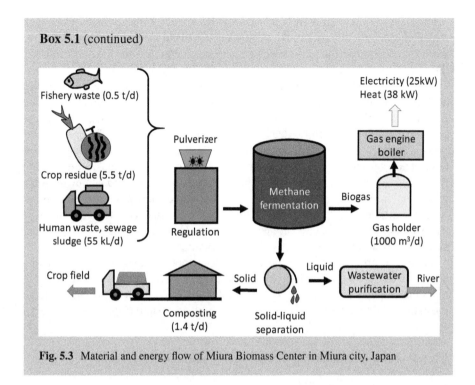

Fig. 5.3 Material and energy flow of Miura Biomass Center in Miura city, Japan

5.2 Production of Combustible Biogas

5.2.1 Production of Biohydrogen

5.2.1.1 Biohydrogen as Prospect Future Energy

Recent energy research emphasizes issues on the depletion of fossil fuel and adverse environmental impacts. The extensive use of fossil fuel causes the emission of a large quantity of waste gas known as the greenhouse gas (GHG). The last decade was the period with the highest GHG emission rate of 2.2% per year (IPCC 2014). This GHG emission is widely recognized as the leading cause of global warming. With the current situation of continuously depleting fossil fuel and its adverse impact on the global environment, renewable energy source has been explored. Using renewable energy is expected to not only provide energy but also solve other problems such as waste discharge, gas emission, and global warming, among others.

Hydrogen gas (H_2) is one of the most common elements on the earth; it is universally recognized as an environmentally safe and renewable energy. The combustion of hydrogen gas produces only water, thus making it an ideal alternative energy to fossil fuels. Hydrogen gas has long been recognized as a potential future energy because its combustion produces a large quantity of energy or 122 kJ/g that is 2.75 times greater than hydrocarbon fuel value (Chandrasekhar et al. 2015). On average, greener energy that causes less adverse environmental impact comes from fuels with higher hydrogen content. Hydrogen fuel has the highest hydrogen content and the least environmental impact when comparing with natural gas, oil, and coal (Dincer and Acar 2015).

Several methods of producing hydrogen gas have been developed: fossil fuel reforming, biofuel reforming, coal and biomass gasification, thermochemical method, water electrolysis (Schmidt et al. 2017), photoelectrochemical method, and biological method. Currently, steam reformation of natural gas is the primary process of producing molecular hydrogen gas. About 50% of hydrogen fuel is from reformation of natural gas, 30% from oil reforming, 18% coal gasification, 3.9% from water electrolysis, and 0.1% from processes. In addition to achieving waste minimization, it recognizes that clean and sustainable energy, tolerance of diverse water conditions, and sustainability, (Dincer and Acar 2015) are critical benefits of using biological methods to produce hydrogen. Hence, this method has a great potential for future development and implementation.

Hydrogen production using the biological method is known as the "biohydrogen" process. It is a natural process, and the hydrogen gas is the by-product of microbial metabolism pathways in biophotolysis, fermentation (photo and dark fermentation), microbial electrolysis, and the combined system (Fig. 5.4).

5.2.1.2   Factors Influencing Biohydrogen Production Operation

Temperature

The influence of temperature on hydrogen fermentation has been widely studied. The optimum temperature condition may vary depending on the type of bacteria growing in the reactor. Currently, numerous reports verify that some bacteria have the capability of producing hydrogen. Based on the optimum temperature for their maximum growth, hydrogen-producing bacteria are categorized into mesophiles (*Clostridium* sp., *Enterobacter* sp., etc.), thermophiles (*Klebsiella* sp., *Thermoanaerobium* sp., etc.), and psychrophiles (*Rahnella* sp., *Polaromonas* sp., etc.) (Wang and Yin 2017). Some other studies use natural mixed bacterial culture obtained from soil, compost, anaerobic sludge, and sewage water, among others as inocula. Because various predominant species of hydrogen-producing bacteria may

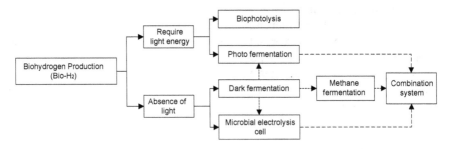

Fig. 5.4 Hydrogen production using biological method

live in different source, the pretreated culture may show a slightly different optimum growth temperature. The mixed culture contains non-hydrogen-producing microbes mixed with hydrogen-producing bacteria. Hence, the raw mixed culture is subject to pretreatment to enrich the desired hydrogen-producing bacteria. Wang and Wan (2008) reported that 40 °C is optimum temperature when using digested sludge as the inoculum. Both 35 °C and 55 °C have been reported to achieve the highest hydrogen yield for bacterial cultures with different dominant microbe species: 35–40 °C for *Clostridium* sp., and 45–60 °C for *Thermobacterium* dominate (Qiu et al. 2017). Hence controlling the reactor at the optimal temperature is of great importance for achieving the maximum hydrogen yield.

pH

Various studies have been carried out to investigate the effect of pH on hydrogen production. The fermentative hydrogen production has various ranges of optimum pH depending on the dominant species living in the reactor. Predominant hydrogen-producing bacterium, *Clostridium* sp., has an optimum range pH of 5.0–6.5. Fang and Liu (2002) conducted an experiment and discovered that the optimal pH is 5.5 when they used mixed culture as the inoculum and glucose as the carbon source. Adjusting pH not only optimizes the growth of hydrogen-producing bacteria but also inhibits methanogenesis microbes. The initial pH, especially for batch fermentation, will possibly affect lag time, synthesis of enzyme, and spore germination, among the many others. Mota et al. (2017) show that when the acid-tolerant *Clostridium* sp. and *Ethanoligenens* sp. are the dominant bacteria, hydrogen is successfully produced under acidic condition (pH < 3.0). Their findings, which contradict the results reported by other researchers, may lead to a new direction of research on the biological production of hydrogen gas.

Nutrient

Macro- and micronutrients are essential to support bacterial growth. Carbon and nitrogen are generally used as the main nutrients for bacterial growth. Utilizing wasted biomass as the substrate could provide the macronutrients required to support hydrogen production. The carbon-nitrogen (C/N) ratio will impact the hydrogen production yield; the optimum ratio will vary depending on sources of substrate and inoculum, for example, the optimum C/N ratio of 47–50 for cattle dung or activated sludge (Mohammadi et al. 2012), whereas C/N ratio of 137 is reported to yield high hydrogen when using synthetic wastewater in an upflow fixed-bed anaerobic reactor (Rojas et al. 2015). Additionally, minerals are micronutrients that enhance hydrogen production, and trace metals are essential for the biohydrogen production. Results of several experiments show enhanced hydrogen yield and bacterial growth with addition of magnesium, calcium, iron (Alshiyab et al. 2008), and nickel (Engliman et al. 2017).

Hydraulic Retention Time

Biohydrogen production occurs at the acidogenesis stage. Extending the fermentation time may shift the bacterial growth from acidogenesis to methanogenesis. Hydraulic retention time (HRT), which is reaction time for microbes to utilize substrate, is of great importance to biohydrogen production. For achieving high hydrogen yield, the HRT needs to be adjusted according to variation in substrate, inoculum, and other operational condition. For instance, the optimum HRTs are 18 h and 12 h, respectively, for petrochemical effluents (Elreedy and Tawfik 2015) and glycerol (Illanes et al. 2017) to reach the maximum hydrogen yield. Operating the reactor at an appropriate HRT is of great importance to achieving a balance between substrate and microbes to get high hydrogen yield.

Partial Pressure of H_2

Hydrogen gas is the target product of biohydrogen fermentation. The microbial hydrogen synthetic pathways are sensitive to the hydrogen produced by the hydrogen production microbes (Levin et al. 2004); increasing H_2 concentration inhibits or reduces the production of H_2 itself. The concentration of hydrogen gas dissolved in the reactor content is directly proportional to hydrogen partial pressure in the reactor headspace. Several studies have been conducted to address the inhibition of hydrogen production by the hydrogen partial pressure. Gas sparging is one strategy to reduce hydrogen partial pressure; sparging the reactor with CO_2, N_2 (Kim et al. 2006), and Ar (Liao et al. 2012) leads to higher hydrogen yield.

Reactor Size

Most current studies on biohydrogen production are carried out using single laboratory-scale reactors. The reactor volume varies from around 100–2000 mL. Understanding the influence of reactor size on biohydrogen fermentation is very important to scale up laboratory reactors for field application. However, few researches investigated the effect of reactor size on biohydrogen fermentation. The experimental results obtained by Alshiyab et al. (2009) show that the hydrogen yield increases with increasing reactor size (67%). Scaling up laboratory reactor to pilot scale (150–500 L) has been successfully done with various hydrogen yield productions (Kim et al. 2010; Cavinato et al. 2012; Lee and Chung 2010). However, scaling up the reactor to an economical size with high hydrogen yield is a challenging task for future studies.

5.2.2 Hot Compressed Water Treatment

Hot compressed water (HCW) is the condition of liquid water when it is subject to elevated temperature and pressure. The HCW temperature is between 100 °C and 374 °C, while the pressure is maintained between 0.1 MPa and 22.1 MPa. The HCW condition is achieved by sparging inert gas into the liquid placed in a stainless container pressurized at a predetermined pressure and heated to reach the subcritical condition (Fig. 5.5).

HCW is a "flexible" or "adjustable" condition; the property of HCW is influenced by its temperature and pressure, and the HCW property is either macroscopic or microscopic. The miscibility, dielectric constant, ionic product, and transport properties in macroscopic view have already been investigated by numerous researchers. Other microscopic factors such as collision frequencies, dipole moment, hydrogen bonds, solutions, and effect on chemical reaction have also been studied (Kruse and Dinjus 2007).

Many agricultural and food industrial wastes contain cellulose or hemicellulose. Under high temperature and pressure, cellulose is degraded into soluble oligomers and levoglucosan (hydrolysis) followed by the degradation of soluble oligomers into glucose and isomers (Buendia-Kandia et al. 2018). On cellobiose thermal decomposition, the low hydrolysis process with isomerization and retro-aldol condensation will occur as dominant primary reactions (Yu et al. 2013). The decomposition behavior of HCW may vary depending on temperature, pressure, and material to be decomposed; it is an important factor to be addressed for applying the HCW condition to treat wasted biomass and recover sugar.

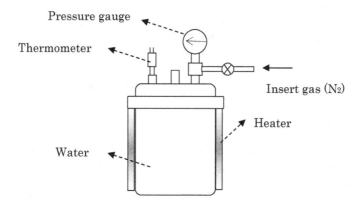

Fig. 5.5 Schematic diagram of the hot compressed water (HCW) device

5.2.3 Biohydrogen Production Using Hot Compressed Water Pretreatment

5.2.3.1 Effect of HCW Pretreatment on Mixed Culture Inoculum

There are many microbes capable of producing hydrogen. Oh and others (Oh et al. 2003; Lee et al. 2010) isolated both aerobic and anaerobic hydrogen-producing bacteria (HPB) such as aerobic *Aeromonas* sp., *Vibrio* sp., *Pseudomonas* sp., *Enterobacter, Citrobacter, Klebsiella, Escherichia coli, and Bacillus* as well as anaerobic *Actinomycetes* sp., *Clostridium* sp., *Porphyromonas* sp., *Ethanoligenens*, and *Desulfovibrio*. In hydrogen fermentation, the mixed culture derived from the environment, wastewater, cattle manure, and others can be used as the inocula. However, the existence of hydrogen-consuming bacteria (HCB) can interfere with the hydrogen production process. Hence, the mixed culture is pretreated to suppress hydrogen-consuming bacteria while enriching hydrogen-producing bacteria in the mixed culture. Many studies have been conducted on using the heat pretreatment by maintaining the temperature below 105 °C to enrich HPB in hydrogen fermentation. This pretreatment basically eliminates nonspore-forming bacteria such as methanogens while enriching spore-forming HPB such as *Clostridium* under extreme condition. When the condition becomes favorable, the spore will then germinate.

Recent studies by Kuribayashi et al. (2017) investigate hydrogen production using the HCW pretreatment. The sludge obtained from a biogas fermentation process is used as the mixed-culture inocula. Contrary to heat pretreatment, the HCW pretreatment operates at temperature above 100 °C under high pressure. During hydrogen fermentation, some *Firmicutes* phylum bacteria becomes dominant in the microbial consortium. The HCW pretreatment suppresses nonspore-forming bacteria. Results of gene analyses reveal that compositions of microbes in heat pretreated and HCW pretreatment inocula are different (Fig. 5.6). *Clostridium* sp. and *Clostridium roseum* dominate in hydrogen fermentation using the HCW-treated biogas sludge as the seeding inoculum.

5.2.3.2 Biohydrogen Production Using HCW Pretreatment

In the biological hydrogen production process, four moles of H_2 is theoretically produced by microbes for each mole of glucose fermented. But the system efficiency of hydrogen fermentation can reach 60–67.7% using pure culture of *Clostridium butyricum* and *Enterobacter aerogenes* (Yokoi et al. 2002; Ueno et al. 2001). However, using the pure culture microbe for hydrogen fermentation has serious limitations in field applications because a strict pure culture condition in a field system is easily contaminated by other non-pure microbial species that will interfere with the growth of the pure culture. Therefore, maintaining a mixed microbe in field application is preferred. Additionally, the mixed culture is broadly available, robust in serious condition, suitable for treating complex materials, and easy to

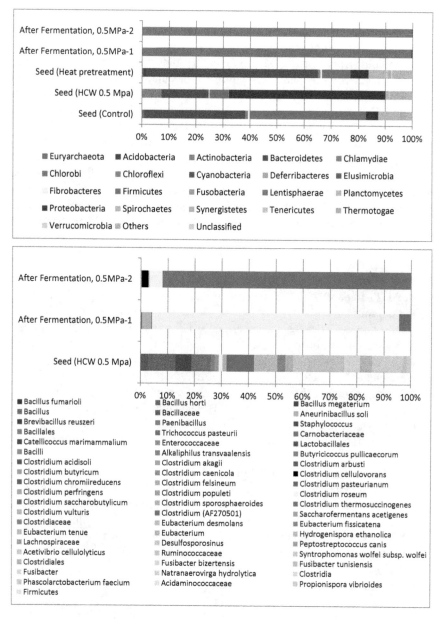

Fig. 5.6 Phylum-level (above) and species-level compositional view from *Firmicutes* phylum (bottom) in hydrogen fermentation using HCW pretreatment

manage. The pretreatment is necessary for promoting the growth of predominant hydrogen-producing species in the mixed culture.

Production of hydrogen using HCW has already been studied using laboratory-scale facilities (Kuribayashi et al. 2017). In this study, the comparison of hydrogen production between heat and HCW pretreated inocula was carried out using about 3.75 g/L glucose as carbon source for microbial metabolism. The result indicates that the HCW pretreatment produces 3 times higher hydrogen gas than the heat pretreatment. The maximum hydrogen yield achieved by using the HCW pretreatment inoculum is 1.5 mol H_2/mol glucose, whereas 0.51 mol H_2/mol glucose is achieved by using the heat pretreatment inoculum. The best HCW pretreatment conditions reported in this study are 150 °C temperature, 0.5 MPa pressure, and 40-min holding time. The dominant bacterial species growing in HCW reactor are *Clostridium* (AF270501 with un-identified species name) and *Clostridium roseum*, both of which can form protective spores under extreme environment temperature (see Fig. 5.6).

Box 5.2 Effect of HCW Pretreatment on the Release of Magnesium and Potassium

Application of HCW (hot compressed water) as pretreatment to promote hydrogen production has been developed recently. The advantages of the HCW are "flexibility" and "adjustability" that makes HCW a potential treatment method; the HCW condition can be adjusted as needed. Biomass is degraded in the HCW treatment to yield ionic products, and some ionic species are essential for bacterial growth. For example, Mg^{2+} and K^+ play important roles as cofactors to stimulate enzyme reactions, as well as phosphate transportation between surrounding environment and cell (Machnicka et al. 2004).

Experiment has been carried out to identify the effect of HCW treatment on biomass associated with ionic product. The HCW experiment runs are carried out based on the combination of two factors: temperature (130, 150, and 180 °C) and pressure (0.5, 1.0, and 1.5 MPa) using dried fine compost as the material. Once the target temperature of each batch treatment is reached, the reaction time starts and continues for 40 min. Ion chromatograph (Shimadzu CDD-6A) is used for analyzing magnesium and potassium. The result shows that magnesium ions are not detected in the untreated sample. After the HCW treatment, the concentration of magnesium ions increases, and the highest concentration of magnesium ions reaches 36.0 mg/L at 180 °C and 1.5 MPa. Potassium ions are detected in the untreated sample (217 mg/L), and their concentration increases after the HCW treatment to reach the highest amount of 264 mg/L at 150 °C and 1.0 MPa (Fig. 5.7).

(continued)

Box 5.2 (continued)

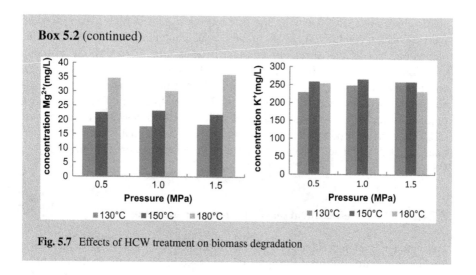

Fig. 5.7   Effects of HCW treatment on biomass degradation

5.2.4   Biohythane Production

5.2.4.1   Two-Phase Fermentation

The biological hydrogen production is widely conducted in many studies. Theoretically, 1 mole of glucose is converted to 4 moles of hydrogen. However, the hydrogen production never reaches this maximum theoretical value in actual practice. Moreover, the biological method of hydrogen production produces acidic liquid rich in organic acids, and the production of the acidic by-product is alleviated by using a two-phase fermentation system. Thus, the shortcomings of producing acidic acids during the production of hydrogen production will be minimized in the two-phase fermentation process. Implementing this method will not only reduce the production of environmentally unfriendly discharge but also increase the energy recovery efficiency.

The two-phase fermentation process consists of two separate reactors. The first reactor is the "dark fermentation" or "biohydrogen fermentation" stage in which hydrolysis and acidogenesis occur. Main fermentation products include H_2, CO_2, and liquids, which are rich organic acids. This phase can be optimized to maximize the hydrogen yield. Many parameters including pH, temperature, HRT, organic loading rate, inoculum, and substrate, among others, have been studied for system optimization. In the second reactor, the organic acid-rich product from the first reactor is used as the substrate to the second reactor where methanogenesis occurs to produce methane, CO_2, and slurry. The methane gas thus produced biologically in the second reactor is called "biomethane." The two-phase system that combines a first and a second reactor to produce respective biohydrogen and biomethane is called "biohythane."

The first and the second reactors are operated under two sets of different conditions so that different predominant microbial species exist in these two reactors. The acidic condition (pH 5.5–6.5) in the first reactor results in high hydrogen yield, whereas neutral condition (pH 7–8) in the second reactor promotes biomethane fermentation. In a single-stage fermentation, it takes around 30 days of hydraulic retention time (HRT) with methane and CO_2 as the main products, whereas two-phase fermentation whole process can be shorter. In first stage, which consist of acidification process, the HRT is less than 3 days with H_2 and CO_2 as main gas product. In second stage, methanogenesis will take 10–15 days of HRT to produce CH_4 as the major product (Hans and Kumar 2018). Additionally, both reactors are operated at different sets of optimal conditions to maximize productions of H_2 and CH_4, respectively.

5.2.4.2   HCW Pretreatment on Biohythane Fermentation

Pretreatment is an important step on the performance of the anaerobic fermentation. Various pretreatments have been applied to get high product yield. In the two-phase fermentation system, pretreatment is a crucial step for the first stage. It is used to decompose complex structure of substrate and suppress hydrogen-consuming bacteria. HCW pretreatment can be used as an alternative pretreatment due to its capability to decompose complex structure such as cellulose (Buendia-Kandia et al. 2018), and cellobiose (Yu et al. 2013), and to suppress hydrogen-consuming bacteria (Kuribayashi et al. 2017). Using the thermal pretreatment will reduce chemical consumptions so that it makes this process more environmentally friendly. The application of HCW pretreatment scheme is shown in Fig. 5.8.

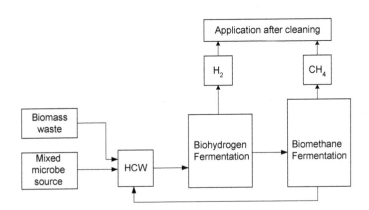

Fig. 5.8 Application of hot compressed water pretreatment on biohythane fermentation

5.3 Electricity Production Using Microbial Fuel Cell

5.3.1 Power Generation by Microbial Fuel Cell

The technological principle of producing electricity directly from biomass using microorganisms has been known for about more than 100 years. During the 1990s, rapid advances in researches on the device and materials led to the development of MFC (microbial fuel cell). Researches on the implementation of MFC are actively carried out in the USA (Logan et al. 2006; Logan and Rabaey 2012) and China (Feng et al. 2010). However, this technology has not been widely implemented for field application primarily because of its low electricity output. The maximum electrical power output is reported to be about 800 mW per m^2 of the electrode area (Miyahara et al. 2016). MFC is expected to be the next-generation innovative technology to generate electricity without causing adverse environmental impacts.

The MFC generates electricity by extracting the electronic current directly from organic matter biologically. However, the biological process to decompose organic matter and produce electricity is time-consuming. As shown by the schematic diagram presented in Fig. 5.9, the integrated MFC system consists of a bio-hydrogen fermentation for producing hydrogen gas as a combustible fuel in the first stage (Kashima et al. 2008) and generating electricity in the secondary stage. The biological hydrogen fermentation in the first stage also generates organic acids that can be used as the substrate for microbial growth in the second-stage MFC to generate electricity. In this process, common organic wastes can be used as the substrate for producing hydrogen fuel and biological electricity. Theoretically, 4 moles of hydrogen and 2 moles of acetic acid, or 4 moles of hydrogen and 1 mole of butyric acid, are produced from 1 mole of glucose as shown in Eqs. (5.1) and (5.2).

$$C_6H_{12}O_6 + 2H_2O \rightarrow 2CH_3COOH + 2CO_2 + 4H_2 \qquad (5.1)$$

$$C_6H_{12}O_6 + H_2O \rightarrow CH_3COCH_3 + 3CO_2 + 4H_2 \qquad (5.2)$$

The MFC process can use acetic acid and butyric acid as substrate for generating electricity following the hydrogen fermentation process.

Fig. 5.9 Bioenergy production using hydrogen fermentation and microbial fuel cell

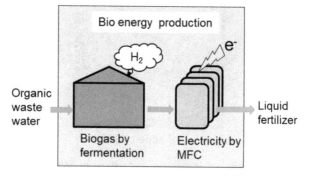

5.3.2 Structure of Microbial Fuel Cell (MFC)

The MFC system has an anode chamber and a cathode chamber partitioned with an ion-exchange membrane. Recently, the single-chamber type of MFC as shown in Fig. 5.10 was developed by integrating the ion-exchange membrane and the cathode electrode. In the anode chamber, the organic matter is biologically decomposed around the anode electrode. The resulting electrons are transmitted out of the anode chamber via the anode. The electron stream moves to the cathode through an external circuit. Once reaching the cathode electrode, the electrons are consumed by oxygen and the proton passing through the ion-exchange membrane from the anode chamber. The electron flow between the electrodes is caused by the difference between the oxidation potential and the reduction potential. The former is generated by the oxidation reaction of the organic substance at the anode electrode, whereas the latter is generated by the reduction reaction at the cathode electrode. Thus, the electron flow passing through the external circuit can be used as an electrical energy.

Besides carbon graphite, other materials such as carbon cloth, carbon felt, and carbon brush are used as the anode electrode. In addition, electrodes with carbon nanotubes are also developed to increase the electrode conductivity. Yamashita et al. (2019) reported that stainless steel oxidized with flame was newly developed as an

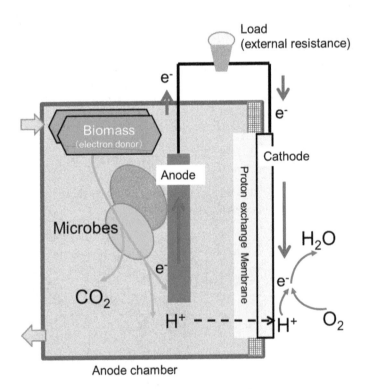

Fig. 5.10 Structure of microbial fuel cell (single chamber type)

anode. Because the electrode supports numerous types of microorganisms that decompose organic matter, the anode material should have a larger specific surface area to accommodate more microbial growth. As the cathode electrode is concerned, a carbon cloth containing platinum is used for easily fixing the organic substances to be oxidized. For the ion-exchange membrane, a membrane commercially known as Nafion is frequently used because of its capability to permeate only proton. Additionally, other types of membrane such as a porous polymer-derived ceramic membrane (PDC, monmorillonite-$H_3PMo_{12}O_{40}/SiO_2$) (Ahilan et al. 2018) that has a high ion-exchange capacity can also be used. Additionally, a PVDF (polyvinylidene difluoride) membrane gripped with ionic liquid has also been developed (Kook et al. 2019).

Iron-reducing bacteria are used as the anaerobic microorganisms; they have high reproduction capability and can form fixed growth easily on the anode electrode. Iron-reducing bacteria acquire energy by decomposing organic materials and giving surplus electrons to the anode, which behaves as an electron acceptor. Both genus *Shewanella* and the genus *Geobacter* are well known to be representative iron-reducing bacteria. These bacteria possess special proteins called "cytochrome c" that plays an important role in transferring the electrons generated inside the cell to the outside medium (Okamoto et al. 2012). In addition, the bacteria produce a special structure that is similar to cilia known as the "electrically conductive nanowire"; this structure plays the role of transmitting electrons to other microorganisms and electrodes among others (Reguera et al. 2005; Lovley and Nevin 2011).

5.3.3 Characteristics of Substrate Consumption and Power Generation in MFC

The power generating capacity, or the power generation density that measures the performance of an MFC, is limited by the areas of both anode and cathode electrodes. Alternatively, the power generation performance is measured by the power generated per unit volume of input substrate.

The electric current value is measured by changing the external resistance when evaluating the power generation capacity of an MFC. Figure 5.11 shows the variation of the voltage generated in a startup MFC experimented in the author's laboratory (Ueda et al. 2016). The output voltage was not stable for the first week; it then became stabilized. The voltage drops abruptly when the acetic acid is consumed completely, and the voltage immediately rises after a new substrate is added to the anode chamber. When 5 mmol/L acetic acid was used as the substrate (PBS-SA), the maximum power density of the laboratory MFC is 450 mW / m^2 that is considerably lower than the power density of other general MFCs. When the residual digesta of hydrogen fermentation is used as a substrate of MFC (HFR), the power density becomes 400 mW/m^2 that is slightly lower than that generated by using acetic acid

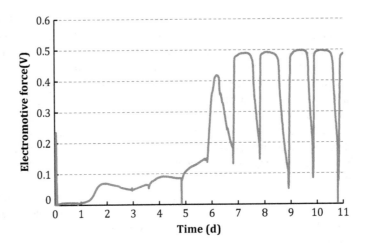

Fig. 5.11 Variation of the electromotive force generated in a startup MFC

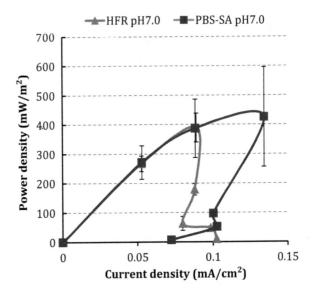

Fig. 5.12 Relationship between current density and power density of MFC

as the substrate (Fig. 5.12). The performance of MFC in the laboratory should be improved more than ten times if the MFC is successful for the real-world application.

As shown by Eq. (5.3), the Coulomb efficiency η is the percentage of the electrons released from the decomposed substrate with respect to the maximum electron charge of the original input substrate.

$$\eta = \frac{\text{Generated electric charge amount by MFC}[\text{C}]}{\text{Consumed electric charge amount}[\text{C}]} \times 100[\%] \qquad (5.3)$$

The following chemical reactions show that hydrolysis of 1 mole of acetic acid results in the production of 8 moles of electrons (Nagatsu et al. 2014).

$$\text{In the anode, } CH_3COO^- + 4H_2O \rightarrow 2HCO_3^- + 9H^+ + 8e^- \quad (E_0 = 0.187 \text{ V}) \qquad (5.4)$$

$$\text{In the cathode, } O_2 + 4H^+ + 4e^- \rightarrow 2H_2O \quad (E_0 = 1.229 \text{ V}) \qquad (5.5)$$

Therefore, up to 8 moles of electrons move from the anode to the cathode. Using the following Eq. (5.6), the charge Q of the 8 moles of electrons is calculated to be 770560 C based on the charge of 1.6×10^{-19} C for one electron.

$$Q = 8 \times 6.02 \times 10^{23} \times 1.6 \times 10^{-19} = 770560 \text{ C} \qquad (5.6)$$

Assuming that this amount of charge moves between the electrodes in 1 day under the potential difference ΔE_0 of 1.2290–0.187 = 1.042 V, the power q of 9.3 W can be obtained, and the corresponding power is about 223 Wh per day.

$$q = \frac{770560}{24 \times 3600} \times 1.042 = 9.3 \text{ W} \qquad (5.7)$$

The voltage loss occurs because some energy is consumed by microbes. The potential difference between the electrodes is about 0.5 V, and the theoretical maximum output is expected to be around 5 W.

5.3.4 Current Challenges and Prospects for Microbial Fuel Cells

Urban agriculture should be managed more strictly with respect to efficiency and environmental issues than rural agriculture because urban regions are more densely populated and have lower buffering capacity than rural regions. The organic waste discharged from the facility for processing agricultural products in urban regions must be well managed, treated, and reused in urban agriculture. The organic garbage discarded from the food manufacturing and restaurants contains a large quantity of moisture. It is suitable to be treated in the anaerobic digestion process such as hydrogen fermentation; the waste slurry can be treated in the MFC process. The combination of these treatment methods is effective in handling the organic waste, as well as controlling its odor efficiently and cost effectively. Most of the electrical and the thermal energy produced from the microbial treatment system will be consumed in operating the system. The digested slurry liquid after the microbial treatment contains abundant nutrients such as nitrogen, phosphorous, and potassium, which are essential to crop growth. And they should be returned to the urban farmland for growing crop. Reuse of the digested liquid may also achieve water reuse in the city.

The MFC system is now undergoing rapid technological innovation and improvement. Some example modifications being developed include using carbon nanotubes with excellent electric conductivity for anode, replacing platinum with less-expensive material for cathode, and modifying the anode chamber structure for enhancing the dominant microorganism to improving the substrate decomposition.

Steady supply of inexpensive power is needed to achieve sustainable operation of greenhouse facilities, agricultural processing, as well as storing and vending facilities. The MFC system is an adaptable technology to meet the power demand needed for operating the above agricultural systems and facilities in urban regions of the world. This will alleviate the dependence on fossil fuels and the emission of greenhouse gases.

References

Ahilan V, Wilhelm M, Rezwan K (2018) Porous polymer derived ceramic (PDC)-montmorillonite-$H_3PMo_{12}O_40/SiO_2$ composite membranes for microbial fuel cell (MFC) application. Ceram Int 44:19191–19199

Alshiyab HS, Kalil MS, Hamid AA, Yusoff WMW (2008) Trace metal effect on hydrogen production using C.acetobutylicum. J Biol Sci 8(1):1–9. Online. https://doi.org/10.3844/ojbsci.2008.1.9

Alshiyab HS, Kalil MS, Hamid AA, Yusoff WMW (2009) Improvement of biohydrogen production under increased the reactor size by C. acetobutylicum NCIMB 13357. Am J Environ Sci 5(1):33–40. https://doi.org/10.3844/ajes.2009.33.40

Buendia-Kandia F, Mauviel G, Guedon E, Rondags E, Petitjean D, Dufour A (2018) Decomposition of cellulose in hot-compressed water: detailed analysis of the products and effect of operating conditions. Energy Fuel 32(4):4127–4138. https://doi.org/10.1021/acs.energyfuels.7b02994

Cavinato C, Giuliano A, Bolzonella D, Pavan P, Cecchi F (2012) Bio-hythane production from food waste by dark fermentation coupled with anaerobic digestion process: a long-term pilot scale experience. Int J Hydrog Energy 37(15):11549–11555. https://doi.org/10.1016/j.ijhydene.2012.03.065

Chandrasekhar K, Lee Y, Lee D (2015) Review: biohydrogen production: strategies to improve process efficiency through microbial routes. Int J Mol Sci 16. https://doi.org/10.3390/ijms16048266

Das S, Bhattacharyya BK (2015) Optimization of municipal solid waste collection and transportation routes. Waste Manag 43:9–18. https://doi.org/10.1016/j.wasman.2015.06.033

Dincer I, Acar C (2015) Review and evaluation of hydrogen production methods for better sustainability. Int J Hydrog Energy 40:11094–11111. https://doi.org/10.1016/j.ijhydene.2014.12.035

Elreedy A, Tawfik A (2015) Effect of hydraulic retention time on hydrogen production from the dark fermentation of petrochemical effluents contaminated with Ethylene Glycol. Energy Procedia 74:1071–1078

Engliman NS, Abdul PM, Wu S, Jahim JM (2017) Influence of iron (II) oxide nanoparticle on biohydrogen production in thermophilic mixed fermentation. Int J Hydrog Energy 42:27482–27493. https://doi.org/10.1016/j.ijhy-dene.2017.05.224

Fang HHP, Liu H (2002) Effect of pH on hydrogen production from glucose by a mixed culture. Bioresour Technol 28(1):87–93. https://doi.org/10.1016/S0960-8524(01)00110-9

Feng Y, Yang Q, Wang X, Logan BE (2010) Treatment of carbon fiber brush anodes for improving power generation in air–cathode microbial fuel cells. J Power Sources 195:1841–1844

Hans M, Kumar S (2018) Biohythane production in two-stage anaerobic digestion system. Int J Hydrog Energy. https://doi.org/10.1016/j.ijhydene.2018.10.022

IEA and FAO (2017) How 2 guide for bioenergy. http://www.iea.org/publications/freepublications/publication/How2GuideforBioenergyRoadmapDevelopmentandImplementation.pdf

Illanes FS, Estela T, Schiappacasse MC, Eric T, Gonzalo R (2017) Impact of hydraulic retention time (HRT) and pH on dark fermentative hydrogen production from glycerol. Energy 141:358–367

IPCC (2014) Climate change 2014: mitigation of climate change. Contribution of working group III to the fifth assessment report of the intergovernmental panel on climate change. Cambridge University Press, Cambridge/New York

Japan Environmental Sanitation Center (JESC) (2012) Solid waste management and recycling technology of Japan: toward a sustainable society. https://www.env.go.jp/en/recycle/smcs/attach/swmrt.pdf

Kashima H, Nyunoya H, Oto M, Tojo S (2008) Effect of bacterial stress on hydrogen fermentation microflora. ASABE Annual International Meeting, Providence, Rhode Island. U.S.A. Paper No.08–4058

Khoiron (2014) Evaluation of municipal solid waste management in Jember District. In: Proceeding of conference: the 5th sustainable future for Human Security (SustaiN 2014). ISSN: 2188–0999

Kim D, Han S, Kim S, Shin H (2006) Effect of gas sparging on continuous fermentative hydrogen production. Int J Hydrog Energy 31(15):2158–2169

Kim DH, Kim SH, Kim KY, Shin HS (2010) Experience of a pilot scale hydrogen-producing anaerobic sequencing batch reactor (ASBR) treating food waste. Int J Hydrog Energy 35(4):1590–1594. https://doi.org/10.1016/j.ijhydene.2009.12.041

Kook L, Kaufer B, Bakonyi P, Rozsenberszki T, Rivera I, Buitron G, Belafi-Bako K, Nemestothy N (2019) Supported ionic liquid membrane based on [bmim][PF$_6$] can be a promising separator to replace Nafion in microbial fuel cells and improve energy recovery: a comparative process evaluation. J Membr Sci 570–571:215–225

Kruse A, Dinjus E (2007) Hot compressed water as reaction medium and reactant properties and synthesis reactions. J Supercrit Fluids 39:362–380

Kuribayashi M, Tojo S, Chosa T, Murayama T, Sasaki K, Kotaka H (2017) Developing a new technology for the two phase methane fermentation sludge recirculation process. Chem Eng Trans 58:475–480. https://doi.org/10.3303/CET1758080

Lee YW, Chung J (2010) Bioproduction of hydrogen from food waste by pilot-scale combined hydrogen/methane fermentation. Int J Hydrog Energy 35(21):11746–11755. https://doi.org/10.1016/j.ijhydene.2010.08.093

Lee H-S, Vermaas WFJ, Rittmann BE (2010) Biological hydrogen production: prospects and challenges. Trends Biotechnol 28(5):262–271

Levin DB, Pitt L, Love M (2004) Biohydrogen production: prospects and limitations to practical application. Int J Hydrog Energy 29(2):173–185

Liao Q, Qu XF, Chen R, Wang YZ, Zhu X, Lee DJ (2012) Improvement of hydrogen production with Rhodopseudomonas palustris CQK-01 by Ar gas sparging. Int J Hydrog Energy 37(20):15443–15449

Logan BE, Rabaey K (2012) Conversion of wastes into bioelectricity and chemicals by using microbial electrochemical technologies. Science 337(6095):686–690

Logan BE, Hamelers B, Rozendal R, Schröder U, Keller J, Freguia S, Aelterman P, Verstraete W, Rabaey K (2006) Microbial fuel cells: methodology and technology. Environ Sci Technol 40(17):5181–5192

Lovley DR, Nevin KP (2011) A shift in the current: new applications and concepts for microbe-electrode electron exchange. Curr Opin Biotechnol 22:441–448

Machnicka A, Suschka J, Grubel K (2004) the importance of potassium and magnesium ions in biological phosphorus removal from wastewater. Polish-Swedish Seminar, Integration and Optimization of Urban Sanitation Systems. https://www.kth.se/polopoly_fs/1.650776!/JPS12p49.pdf

Miyahara M, Hashimoto K, Watanabe K (2016) Use of cassette-electrode microbial fuel cell for wastewater treatment. J Biosci Bioeng 115(2):176–181

Mohammadi P, Ibrahim S, Annuar MSM, Ghafari S, Vikineswary S, Zinatizadeh AA (2012) Influences of environmental and operational factors on dark fermentative hydrogen production: a review. Clean (Weinh) 40(11):1297–1305. https://doi.org/10.1002/clen.201100007

Mota VT, Júnior F, Trably E, Zaiat M (2017) Biohydrogen production at pH below 3.0: is it possible? Water Res 128:350–361. https://doi.org/10.1016/J.WATRES.2017.10.060

Nagatsu Y, Tachiuchi K, Narong T, Hibino T (2014) Factors for improving the performance of sediment microbial fuel cell. J Jpn Soc Civ Eng (B2) 70(2):1066–1070

Oh YK, Park MS, Seol EH, Lee SJ, Park S (2003) Isolation of hydrogen-producing bacteria from granular sludge of an upflow anaerobic sludge blanket reactor. Biotechnol Bioprocess Eng 8(1):54–57

Okamoto A, Hashimoto K, Nakamura R (2012) Long-range electron conduction of Shewanella biofilms mediated by outer membrane C-type cytochromes. Bioelectrochemistry 85:61–65

Qiu C, Yuan P, Sun L, Wang S, Lo S, Zhang D (2017) Effect of fermentation temperature on hydrogen production from xylose and the succession of hydrogen-producing microflora. J Chem Technol Biotechnol 92(8):1990–1997. https://doi.org/10.1002/jctb.5190

Ramos TRP, Morais CS, Barbosa-Póvoa AP (2018) The smart waste collection routing problem: alternative operational management approaches. Expert Syst Appl 103:146–158. https://doi.org/10.1016/j.eswa.2018.03.001

Reguera G, McCarthy KD, Mehta T, Nicoll JS, Tuominen MT, Lovley DR (2005) Extracellular electron transfer via microbial nanowires. Nature 435:1098–1101

Rojas MPA, Fonseca SG, Silva CC, Oliveira VM, Zaiata M (2015) The use of the carbon/nitrogen ratio and specific organic loading rate as tools for improving biohydrogen production in fixed-bed reactors. Biotechnol Rep 5:46–54. https://doi.org/10.1016/j.btre.2014.10.010

Schmidt O, Gambhir A, Staffell I, Hawkes A, Nelson J, Few S (2017) Future cost and performance of water electrolysis: an expert elicitation study. Int J Hydrog Energy 42(52):30470–30492. https://doi.org/10.1016/j.ijhydene.2017.10.045

Ueda M, Ichioka T, Chosa T, Tojo S (2016) Electricity generation characteristics of microbial fuel cell using substrate of volatile fatty acids from hydrogen fermentation. ISMAB 2016, Niigata, Japan, IV-13

Ueno Y, Haruta S, Ishii M, Igarashi Y (2001) Characterization of a microorganism isolated from the effluent of hydrogen fermentation by microflora. J Biosci Bioeng 92(4):397–400. https://doi.org/10.1016/S1389-1723(01)80247-4

United Nations Environment Programme (UNEP) (2005) Solid waste management. CalRecovery Inc. www.unep.or.jp/ietc/publications/spc/solid_waste_management/Vol_I/Binder1.pdf

Wang J, Wan W (2008) Effect of temperature on fermentative hydrogen production by mixed cultures. Int J Hydrog Energy 33(20):5392–5397. https://doi.org/10.1016/j.ijhydene.2008.07.010

Wang J, Yin Y (2017) Biohydrogen production from organic wastes. Springer, Singapore. https://doi.org/10.1007/978-981-10-4675-9

Yamashita T, Ishida M, Asakawa S, Kanamori H, Sasaki H, Ogino A, Katayose Y, Hatta T, Yokoyama H (2019) Enhanced electrical power generation using flame-oxidized stainless steel anode in microbial fuel cells and the anodic community structure. Biotechnol Biofuels 9:62

Yokoi H, Maki R, Hirose J, Hayashi S (2002) Microbial production of H_2 from starch-manufacturing wastes. Biomass Bioenergy 22(5):389–395. https://doi.org/10.1016/S0961-9534(02)00014-4

Yu Y, Shafie ZM, Wu H (2013) Cellobiose decomposition in hot-compressed water: importance of isomerization reactions. Ind Eng Chem Res 52(47):17006–17014. https://doi.org/10.1021/ie403140q

Chapter 6
New Technologies to Implement Precise Management of Farming in a City

Tadashi Chosa, Hitoshi Kato, and Rei Kikuchi

Abstract Robotics is a key technology to innovate various industries such as manufacturing, medical care, nursing care, transportation, and agriculture. The robot that can talk and communicate with people is currently available on the market. Robot tractors have been studied for a long time since the 1990s. The accumulation of technology has assisted recent ICT agricultural and robotic technology. Applications of drones to monitor crop growth, the growth environment, and application of materials, among the many others, have increased, and the deployment of flying robots is expected to expand. The situation that miniaturization of sensors, progress of AI, and dissemination of communication technology, etc. are progressing daily favors the development of small robots. The use of compact robots in housework, logistics, medical care, and the like that are close to our daily living just starts. Application of small-sized robots in actual agricultural work and contribution to agricultural production is discussed in the first section.

The Global Navigation Satellite System (GNSS) is a generic term for the generally recognized GPS, QZSS, and other satellite positioning systems. Calculation algorithm of positioning system is described in the second section to realize a driving support system for a farming guidance system based on GNSS and automatic driving assistance.

Producing fruits around urban regions is popular in Japan. However, the fruit production around a city has some issues that need to be addressed. When the farmers spray pesticides and apply chemical fertilizers in the field that is close to houses,

T. Chosa (✉)
Institute of Agriculture, Tokyo University of Agriculture and Technology, Fuchu, Tokyo, Japan
e-mail: chosa@cc.tuat.ac.jp

H. Kato
Division of Lowland Farming, Central Region Agricultural Research Center, National Agriculture and Food Research Organization, Tsukuba, Japan
e-mail: katojin@affrc.go.jp

R. Kikuchi
Division of Farming Systems Research, Western Region Agricultural Research Center, National Agriculture and Food Research Organization, Fukuyama, Japan
e-mail: kikuchi_r@affrc.go.jp

© Springer Nature Singapore Pte Ltd. 2020
S. Tojo (ed.), *Recycle Based Organic Agriculture in a City*,
https://doi.org/10.1007/978-981-32-9872-9_6

the residents will complain about the work. Managing work on fruit production should be done precisely and carefully in urban region. The various information from a plant body such as leaves, branches, flowers, etc. provides management solutions on irrigation timing, amount of water to farmers, etc. The information includes changes in photosynthetic capacity, transpiration, and water transport inside a plant body, among the many others. An attempt monitoring the stem diameter displacement of blueberry plant by using a small load cell with high precision is explained for obtaining biological information in the third section.

6.1 Advanced Agriculture Using a Small Robot

6.1.1 Robotizing Agriculture and Disadvantages of the New Technology

"New Robot Strategy in Japan" was released by the Headquarters for Japan's Economic Revitalization in 2014. Japanese government realizes that robotics is a key technology to innovate various industries such as manufacturing, medical care, nursing care, transportation, and agriculture. Recently, people are not surprised to see a bipedal walking robot, a cleaning robot, or a robot with a smart speaker to help with our ordinary daily living. The robot that can talk and communicate with people is currently available on the market. Humanoid robots with receptive responses are used in some hotels. The delivery service provided by unmanned aerial vehicles has been experimented as a demonstration project. Implementing the automatic technology for driving cars is also in progress. In Japan during the 1980s, robot technology is realized to enhance higher productivity. Nowadays, these robotic technologies have become a familiar way of living in our lives.

2018 is said to be the first year of robot agriculture in Japan. Full-scale tractors with automatic functional features became available commercially, and the Ministry of Agriculture, Forestry and Fisheries (MAFF) promulgated guidelines on tractor safety. Robot tractors have been studied for a long time, and by the 1990s, the practical robotic technologies were presented (e.g., Yukumoto et al. 1998). The accumulation of technology since that time has assisted recent ICT (information and communications technology) agricultural and robotic technology in accelerating the practical use of automatic machinery in addition to improving the surrounding environment such as the formulation of safety guidelines on using the automatic machinery. Also, the issue on expansion of agricultural plot scale and lack of alternatives to substitute tractors cannot be overlooked. Use of robot tractors is becoming reality for the agricultural industry in Japan. Techniques for the autonomous traveling of rice planting machines (Nagasaka et al. 1999), combined harvesters (Saito et al. 2012), and tractors have been reported for a long time; they are expected to be put into practical application. In a few decades, an integrated mechanization system will be developed. The rice cultivation in Japan has reached the stage of considering the use of a robotic system. These technologies contribute to improving productivity, alleviating labor requirements, and supporting large-scale management in a nation with issues of aging population.

The large-scale management of robotic agriculture will probably not be deployed in all regions. While scale expansion of agricultural plots is pursued to improve productivity, expanding the cultivation of abandoned land may not be practical for disadvantaged places such as the intermountainous regions. This is because farmland conservation, flood control, and disaster prevention need to be implemented to protect wildlife. Issues such as population aging, lack of manpower, and expansion of cultivated abandoned areas accelerate depopulation, which is also a factor resulting in the collapse of a region. These problems cannot be solved by using robotic agriculture alone. Agricultural production and revitalization of local industry in disadvantaged regions are also important issues. Robots are effective in the development of agricultural machinery and agricultural techniques. In this chapter, we focus on production systems with small robots and discuss their advantages and challenges in addition to introducing the development of related technologies.

Box 6.1 Will All Agricultural Work Be Conducted by Using Robots in the Future?

The writer's university offers a seminar discussing the robotization of agriculture to freshmen students. In addition to the writer, the panel includes six freshmen and one moderator who studies small agricultural robots in a master's program. The seminar was offered ten times each semester in a course entitled "Basic Seminar of Agricultural Science" at Tokyo University of Agriculture and Technology in 2018. The theme of the seminar was "Can Robots Do Agriculture?" People are not always familiar with both agriculture and robotics. The writer was interested in studying the view of agricultural robotizing by consumers who are not engaged in agricultural engineering nor agricultural production.

First, the panel had a discussion on the current status of agriculture. The class members studied statistics, newspaper articles, and web information, among other sources of information. They then presented the results of their investigation and shared the understanding of future problems in agriculture. The next topics covered robotization in several situations such as industries, logistics, domestic chores, medical affairs, etc. After the seminars, the students have learned that robots will play an important role in solving the problems of population aging, lack of manpower, and rural devastation. An interesting point that has been pointed out is that there is a sense of incongruity for using robots to carry out all the work. The writer feels that they seemed to have a connection for food, food education, and life as a concept derived from agriculture. The food production system is a type of process for achieving a sustained use of plants and animals, and human beings are active participants in this system to enhance the advancement of this production process. The robotization of agriculture will eventually be developed so that mankind will remain a member of and manage the agricultural systems.

6.1.2 Limit of Plot Scaling Agricultural Productivity

The autonomous driving is a practical technology for operating large riding-type agricultural machines, and its future progress is expected to support large-scale management of agricultural systems. Meanwhile, the improvement of productivity that depends on the use of large-scale machinery and applying massive input of materials has reached a limit. Figure 6.1 shows the relationship between the business scale of paddy rice in Japan and the production cost per unit area. By expanding the scale of management, the production costs per unit area are reduced. With a management scale of less than 0.5 ha (hectare), the production cost per ha is more than 1,800,000 Japanese yen, but by expanding the scale to 5 ha, the production cost per unit area is reduced by half. According to the statistics published for 2010, the average size of a lot is less than 1.5 ha so that the production cost can be lowered simply by promoting scale expansion of agricultural lots in the future.

However, the rate of cost reduction becomes moderate for larger plot scale, but the ultimate cost reduction does not reach zero. For agricultural corporations conducting advanced management, the scale of management is commonly over 10 ha. For such a management scale, the production cost may not be significantly reduced with a further expansion of the scale. In domestic arable agriculture, implementing scale expansion and robotized mechanization in the future will reduce the working hours per unit area and increase the profit. It can be said that we are facing a new situation in which the profit may not increase significantly even if we expand the plot scale and develop robotic operations accordingly. In the United States, the production cost for an average plot of around 300 ha is about $4000 USD per ha, which cannot be achieved by just magnifying the scale.

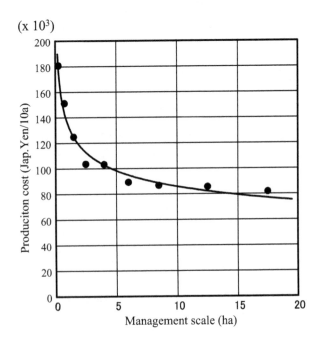

Fig. 6.1 Reduction of rice production cost with respect to management scale in Japan. (Data were cited from eStat; a portal site for Japanese Government Statistics)

Will using a small robot be an alternative method to resolve this situation? Organized sessions probing the possibility of applying small robots in agricultural production have been planned every year since 2016 at the annual meeting of the Japanese Society of Agricultural Machinery and Food Engineers (e.g., Chosa 2017). Discussions center on the effectiveness of small robots, organic farming methods, and affinity with renewable energy for various crops.

6.1.3 Advantages and Disadvantages of Small Robots

Small robots that travel autonomously around crops are often visualized as agricultural technologies of the future because they have not been realized yet. Small machines that are excellent in portability and mobility will be significantly cheaper and safer than conventional riding-type agricultural machines. Hence, many efforts have been attempted to introduce small robots into agricultural applications (such as Pedersen 2006). In recent years, applications of drones to monitor crop growth, the growth environment, and application of materials, among the many others, have increased, and the deployment of flying robots are expected to expand. However, because the net output is also reduced due to increasing cost to miniaturize the machine, the work that can be handled by using robots is practically limited to monitoring crop growth and the growing environment, weeding, and similar operation for the time being. The situation that miniaturization of sensors, progress of AI, dissemination of communication technology, etc. are progressing daily favors the development of small robots. The use of compact robots in housework, logistics, medical care, and the like that are close to our daily living just starts. However, as the application of small-sized robots in actual agricultural work is concerned, the output is not enhanced; thus, small-sized robots are considered not contributing to agricultural production directly.

Specifications of Japanese rice transplanters are investigated; the data excerpted from information published in brochures and websites are plotted in Fig. 6.2. The figure shows how the efficiency is affected by upsizing the machine by using the Japanese rice transplanters as an example. The left ordinate is the "theoretical field capacity" that is calculated based on the machine operational width and maximum operating speed. The right ordinate shows the price divided by engine power consumption. Higher engine output increases the theoretical workload as shown by the solid line with black circle data points in the figure. The machine working width is widened to accommodate increasing machine size, and the greater machine width is possible to enhance the work speed. The efficiency of agricultural operations depends on prompt and timely completion of the work during an appropriate time period favorable to agricultural operations. Completing work for a large plot area in a limited time is of great advantage. Although the amount of work is proportional to the size of plot, multiple machines can be used to handle the extra work; several operators are needed to operate individual machines. If a large machine with the horsepower equivalent to the total horsepower of the multiple machines is used, the number of operators can be reduced to lower the operating cost. Further, as the cost of the machine is considered, the initial cost increases for using machines with

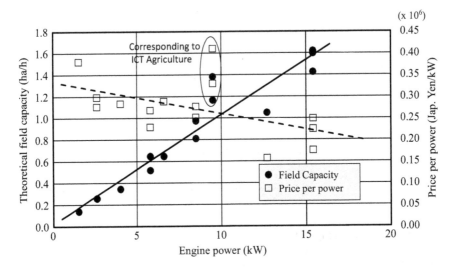

Fig. 6.2 Theoretical field capacity and price per unit power for Japanese rice transplanter vs. engine power

larger engines or more power output. A slightly different tendency is seen when calculating the price per engine output. For example, the price per engine output of a 2.1 kW rice transplant machine exceeds 0.35 yen/kW. When the engine output exceeds 15 kW, the price per engine is reduced to less than 0.25 yen/kW. Two models of the 9.5 kW machines are more expensive than other models, but these two models have an information communication function to correspond with cloud services. The 7.7 kW model is particularly less expensive because the engine is mounted conforming to the working width. There is an eight-row planting model with 9.5 kW output; the applicable model is the six-row planting type. This model is inferior to the other models in terms of field workload, but the engine has a useful power output. As described above, although the price is influenced by additional values for larger and more powerful machines, and the user's preferences, in general, the price per engine output tends to decrease as the engine output increases as shown by the dotted line with white square data points. In conclusion, using large machines is shown to reduce the agricultural production cost.

A general perception is that small machines are disadvantageous in terms of agricultural work efficiency and production, but will small robotized machines that work autonomously improve the situation? The robotized small machines perform the same work as a large machine when the equivalent total power is concerned, but the work is shared by multiple machines without increasing the number of operators. In addition, unlike human operators, the robotized machines can work continuously regardless the time of the day, and hence the efficiency is not a concern. For large robotic machines, an accident such as colliding with each other may cause serious consequences on property loss and human lives, and hence, the safety specifications must meet high standards. However, if smaller robot machines are used, the damage caused by accidents may not be as serious as that caused by using large robot machine. Hence, for small robot machines, safety standards become compromised when comparing with large robotic machines.

In response to the myth that agricultural output is reduced by using miniaturized machines, the writer and colleagues have demonstrated that robotization of small agricultural machines does not suffer from a reduction in efficiency. There is a concern that the work may be limited concerning the output reduction. In agricultural production, as with other industrial products, the size of a single device is not reduced just because of a smaller device size. Even if the machine is miniaturized, the ensemble does not become smaller until the final agricultural products are produced. In addition, the plot size and water consumption in the production process are not related to the machine size, but the necessity of transporting and handling heavy objects is still unchanged. As a solution to the transportation issue, Soga has started a new farming method by developing rice transplanters and inventing seedling mats. In addition, transplantation of seedling mats and young seedlings becomes popular as a result of the advances of agricultural mechanization. Application of agricultural mechanization for managing fertilizer has been studied. Furthermore, grain harvesting is done by reaping, sun drying, and threshing, but the order of threshing and drying is reversed with the appearance of the combined rice harvester. Grain shedding in this process is also an important consideration in breed improvement. The incompatibility of using small robots with the existing agricultural system must be addressed by adapting the cultivation methods to small robots. However, rather than waiting for the development of a whole new agricultural method, Chosa (2017 and 2018) suggests to evaluate the many available agricultural methods that have potential to adapt to small robots.

Using the non-tillage cultivation as an example, it is literally a type of cropping system by skipping tillage, and this practice has been studied for a long time (e.g., Faulkner 1943). Substantial labor saving can be achieved by not cultivating the soil during the agricultural process. This method may result in problems such as unstable germination, insufficient root elongation, and difficult weed control, which are thought to occur during the early stage of tillage cultivation. Some reports show that these problems are alleviated after long-term successive use of the uncultivated land, and both growth and yield are equivalent to those using customary cultivation (Komatsuzaki 2007). A system that can simplify or eliminate the soil handling, which requires high power for small robot operation, is very attractive. For the current situation, the soil handlers may be enhanced by adding a small robotic component. Many technologies such as organic cultivation, crop covering, and direct seeding are expected to be compatible with small robots. Organic cultivation requires efficient management of weeds and control of insect pests without applying chemical agents. In a system with large machines, processing crops into food is rather difficult, but the circumstances change completely when small robots are used to do most of the work. Without robotic machines, sowing seeds at the same time under varying soil conditions is expected; this may lead to uneven germination. Using small robotic machines, sowing seeds can be carried out in accordance with locally changing soil conditions. There are many different viewpoints on labor saving and environmental conservation. Sometimes, the problems associated with these systems are not caused by technologies themselves but inefficient management of the agricultural field and/or large farm machinery.

Prof. Simon Blackmore of Harper Adams University pointed out in his publications and lectures some limitations of the advancement of large-scale agricultural machinery and suggests that small agricultural machines be developed (e.g., Blackmore 2016a, 2016b). According to Blackmore, agricultural machinery continues to grow in size that causes increasing scale of the business of the agricultural industry. However, very few agricultural producers can use agricultural machinery of larger sizes. For example, a large tractor is too clumsy to turn around in a general field so that it can only be used in certain specific fields. In addition to issues on handling machinery, the soil tolerance to large and heavy tractors is also a limiting factor to using larger agricultural machines. Therefore, a large tractor does not always operate at its full capacity to yield the maximum output. Furthermore, a tractor is generally used for various agricultural operations such as tilling, seeding, spraying materials, cutting, and transporting, among others. The required machine output for achieving these tasks is not uniform. A large tractor prepared for high-output operations such as cultivation and transportation may not be suitable for carrying out seeding and material scattering operations in addition to raising the problem of soil compaction. After the soil is compacted by large tractor traveling through the field for a long time, the growth of the crop's roots is deteriorated. Increasing the size of the tractor further strengthens the soil compaction so that a larger tillage power is required to cultivate the strongly compacted soil. As the tractor becomes larger, the fuel consumption for powering the tractor also increases.

Box 6.2 Is a High-Precision GNSS Required for the Control of Small Robot? (1. Accuracy vs. Cost of the Navigation System)

Until the 1980s, numerous position recognition techniques in the field had been reported using internal sensors and various navigation devices. However, with the popularization of civilian GPS (Global Positioning System) receivers in the latter half of the 1990s, GPS became the major position recognition technology for autonomous driving of agricultural vehicles. With disablement of SA (selective availability), the operation of QZSS (Quasi-Zenith Satellite System), etc., the development of agricultural robots has advanced significantly. As a result, a robotic rice transplanter controlled by using RTK (real-time kinematic)-GPS is more precisely maintained, even for one stock and one-hand planting operation. GNSS (Global Navigation Satellite System) is a common term nowadays; even though the GPS was originally developed and managed by the United States, it will be an important navigation system for field robotic machines. The IMU (inertial measurement unit) is also a vital device.

There are many technical and economic advantages of applying a highly accurate navigation supporting system; cost is one major advantage for small robots. Internal sensors and a two-dimensional distance meter are the main devices for position recognition used in a household cleaning robot. Further, the lawn-mowing robot recognizes the boundary by the perimeter, but it does not have a rangefinder. Although it uses GNSS for position recognition, the receiver is a low-cost standalone system for auxiliary use. Car navigation systems commonly adopt GPS as an auxiliary system in automobile driving. Recently, the autonomous driving of agricultural machinery is highly dependent on GNSS, however, this control technology is free from expensive GNSS or IMS may be required low-cost robots.

Box 6.3 Is a High-Precision GNSS Required for the Control of a Small Robot? (2. Simulating Seeding That Does Not Depend on a Navigation System)

If the seeding work is carried out by random running within a predetermined range, a robotic seeding distribution machine with reduced bias is shown to be possible. When the seeding robot sows individual seed grains, the seeding distance is determined by the traveling speed of the robot and the time interval of dropping seeds. Fujimori et al. (2017a) reports that when the seeding interval is short, the dispersion of the seeding distribution increases. Increasing the seeding interval will reduce the seeding variations. When the seeding interval exceeds 1 m, it converges to a certain range. Furthermore, the seeding distribution is alleviated by not seeding at turning of the field. Figure 6.3 shows that the area occupied by one seed is indicated by the Voronoi diagram, and the seeding distribution is represented by the degree of the variation. The convergence of variation becomes fast by adopting spiral running instead of complete random running (Fujimori et al. 2017b). A control that combines random running, pattern running, and others is also adopted for cleaning robots and lawn-mowing robots. Fujimori's report shows a control that does not depend on using expensive GPS.

Fig. 6.3 Evaluation of seeding distribution. The seeding distribution was evaluated by the standard deviation of the area in the Voronoi polygon. The value decreases and converges with increasing seeding distance by random traveling. The results obtained before and after revision are shown

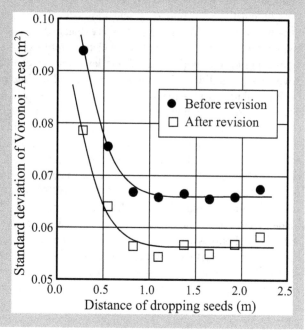

6.1.4   Practical and Experimental Attempts of a Small Robot in Agriculture

Until now, the scale of agricultural management and the size of agricultural machinery have been shown to reach their respective limit thus leading to the tendency of using small agricultural machines. If judged based on the global standards, Japanese agricultural machinery is considered small enough. The miniaturization of agricultural machinery was stated as a novel technology to overcome the limits of enlargement and realize innovation in agriculture. A miniaturized machine is defined as the machine having about 30 kg in mass and external dimension of less than 1 m that can be operated by a single person. The capacity of such miniaturized machines to perform the farm work is shown below.

6.1.4.1   Robotic Lawn Mower

The robotic lawn mower is a well-known outdoor robot that has already become popular. In Europe and the United States, many manufacturers have released various models for household use. Many models employ a perimeter technique in which boundary wires with a weak current flowing through the wire is used. The robot runs within the wire boundary and mows the lawn. The robot is powered by a lithium-ion battery. When the battery is exhausted, the robot returns to the charging station for recharging the battery and then returns to work after fully charged. Depending on the model, the robotic lawn mower can be programed to mow the lawn on specified dates in advance with regard to time of the day and week. The operational noise of the machine is extremely low so that it can be operated at night. Therefore, once the boundary cable is embedded and connected as well as the charging station and the power supply provided, the robot will mow the turf in a semipermanent, maintenance-free manner. Kato et al. (2016) tried to cut weeds other than turf with this robot for managing a bare land. The lawn-mowing robot was installed in a 10 m × 6 m section of the bare field and left there for several months. The height of the model (AUTOMOWER® TM330X, Husqvarna) in use was 308 mm, whereas the weed height was more than 1 m. Because the cutting blade is fixed (not rigid) to the edge of the disk under the main body, grass cutting is performed only when the robot traverses. The grass with soft stems can be cut by the robot on the first pass, whereas the densely growing grass may have stiff or soft stems. The robot makes decision on whether the vegetation growth is an obstacle or not; it will bypass the growth after a "collision." However, repeated collisions may push down the grass with stiff stems; the robot will cut the grass on the subsequent pass. As a result, except for the lignified grass, the mower will complete mowing the weeds. In a trial test, the writer and colleague completed the weeding in a test area by running the robot for about 45 h in 9 days. Once the area was mowed, the robot continued future mowing operations smoothly without considering the grass as an obstacle. In normal mowing operations, collecting and recovering the mowed grass are necessary subsequent operations. If the grass is mowed by using this machine, it is cut into fine pieces bit by bit. Normally, a mower cuts grass in a short time for a couple of times a year, and the grass grows between two mowings. However, if the grass is mowed by more frequent intermittent cuts with a lawn mowing robot, there is no sufficient period for the grass to grow so that this practice changes the vegetation growth. However, Kato et al. (2016) showed that such introduction of robotic mowers has no influence on the growth of grasses such as *Plantago lanceolata* that develop laterally, whereas the growth of grasses such as *Agropyron kamoji* that develop vertically was suppressed. Cases where intermittent grass cutting by robot mowers was effective for suppressing *Solanum carolinense* are reported.

6.1.4.2   Air-Assisted Seeding Machine

The air-assisted seeding machine (Chosa et al. 2009) transports seeds by using the suction of a lateral air within a wide working width of about 10 m. This causes the seeds to fall with the force of air that blows toward the field surface. This machine was developed for direct seeding of paddy rice to form strips and to reach a certain degree of seeding depth. The seeding machine has about 20–40 kW. To work flexibly and efficiently on-site under field conditions such as the maintenance of farm road turns, etc., 2 ha of field work per hour is needed. Indeed, the air-assisted seeding technology supports large-scale operations. Attempts have been made to adapt a similar seeding technique to a small robot (Chosa et al. 2015, Chosa 2018). The prototype has a simple mechanism with an area of about 10 mm × 10 mm that allows the float to be fixed to the lower part of the plot farm seeder, whereas the power spreader is attached to the upper part. The power spreader can spread the seed over a wider range by using a spraying tube that sways the boom from left to right and back alternatively via a T-shaped pipe. Holes are drilled at even intervals in the spray pipe with a collision plate being attached to each hole; seeds collide with the collision plate so that they fall through the hole. Dropping seeds evenly from each hole of the sparging tube and conveying irregularly shaped particles with a single tube are not easy tasks. Some devices are needed to arrange the collision board. No such self-propelled models have been in previous report, but the controlled running is made possible by including a running function in the software. It is assumed that the device will be used in soil underwater. When the water depth in the rice paddy is shallow, the machine will move along while dragging the field bottom. If the water is deep enough, the machine will float and move easily. Therefore, for the machine to travel effectively in the paddy with varying water depth, a mechanism that enables amphibious traveling is needed. For a tractor-mounted large machine, shortening the distance between the spray pipe and the field surface is difficult. However, by reducing the size of machine, the spray pipe can be arranged at a position close to the field surface. Therefore, even if the soil condition and water depth change, the device can be adjusted so that a more appropriate seeding depth can be obtained.

6.1.4.3 Energy Supply to Small Robots

Machine miniaturization has also some advantages regardless of the negative effectiveness caused by machine miniaturization as mentioned in above sections. When the machine output is smaller, the machine consumes less energy. Not all machinery used in agricultural operations requires a high yield. For the lawn mowing and other operations as mentioned in above sections, the required output is low as for a small task. If renewable energy is used for agricultural machinery, developing new technologies is required to modify the machinery. For a small lawn-mowing robot, the needed new technology for using renewable energy can be easily developed and implemented. Renewable energy such as sunlight, wind power, geothermal power, small- and medium-sized hydropower, and biomass, among others, become attractive from the viewpoint of reducing CO_2 emission and achieving energy security. However, renewable energy has the disadvantage of being unreliable due to the technical issues associated with the storage of the electricity generated. If solar energy is used to power miniaturized and robotized agricultural machines, the working efficiency is affected by the supply of the energy source. However, some small robotic agricultural machines are permissible to work slowly regardless of day or night depending on the situation and environment. Many case studies have been reported on using a standalone solar panel to power the abovementioned robotic lawn mower (Kuroda 2017). The standalone power supply consists of a solar panel plus a charge controller, a lead storage battery, and an AC/DC converter. Solar energy charges the lead storage battery during the day, and the battery supplies the current when the robot returns to the charge station. Future tasks include optimizing the efficiency of the robot and time to charge the battery according to the season and the weather, as well as devising a method to manage a larger area by examining the energy use efficiency and the machine scale, among others. However, the device itself serves as an example of practical machine to make it possible to introduce the lawn-mowing robot to locations where commercial power supply is not available. The compatibility between small robotic agricultural machines and renewable solar energy is exemplified in this case.

There are not many renewable energy methods being practically used for the time being; however, research on the use of renewable energy is continuously progressing to search for energy sources and improving the conversion efficiency. Sometimes, the direction of technology development is influenced by changes of social conditions and policies that may impede the implement of a practical technology. Wide, thin, but enormous energy sources remain unused in the present situation. If energy sources are wide and thin, they may be more attractive for current use than focusing on the technological development of energy conversion. The possibility of using a small robot can be recognized from the viewpoint of energy utilization technology rather than technology development to support mass production. Changing the entire social system is difficult so that it is currently impossible to establish a production system that does not depend on fossil fuels at least for agricultural production.

Box 6.4 Future Mobility of a Small Robot

The scene of a young man running around the town with a flying skateboard is seen in a 2015 movie. Although the flying skateboard is not realized at this moment, a variety of vehicles discussed in the following sections can be considered as a flying skateboard. CuboRex Co., Ltd. (Nagaoka, Japan. President; Mizuhito Terashima) has developed an electric crawler extension kit "CuBoard" allowing the crawler to run on gravel roads, snowy roads, and lawns with skateboards (Fig. 6.4). The size of the crawler unit is 320 mm × 240 mm × 100 mm, and it weighs 3.5 kg with a maximum speed of 20 km/h. The kit that can be charged from a 100-V household electric outlet has a range of 10 km for 3 h after it is fully charged. It can be modified into an electric skateboard, electrically assisted carry pack, motorized bicycle, etc., and load capacity of the modified vehicle is 100 kg. Terashima is marketing the modified kit as the world's smallest snow car that is also available for running on public roads.

This kit is not specially developed for an agricultural robot, but it has many of the necessary functions of small field robots to run on rough dirt roads and move loads. Its high durability, low weight, and small size are useful characteristics for agricultural robots. The development of new mobility technology will boost the use of small robots in agriculture.

Fig. 6.4 Running CuBoard which is carrying a person at a rice paddy, after harvesting

6.2 Agricultural Technology Using GNSS

6.2.1 What Is GNSS

6.2.1.1 Mechanism of GNSS Positioning

The Global Navigation Satellite System (GNSS) is a generic term for the generally recognized GPS, GLONASS, Galileo, QZSS, and other satellite positioning systems. GNSS positioning methods include standalone positioning performed by using one receiver and relative positioning performed by using two or more receivers. The terrestrial receiver receives radio signals transmitted from multiple artificial stationery satellites. By measuring the differences in the slight delay of the radio waves from respective satellites, the receiver calculates the coordinates of its current position on the ground. Three satellites can give the coordinates by using the triangulation method. However, since the clock built into a receiver is not extremely accurate, the distance from a fourth satellite is needed to improve the accuracy (Fig. 6.5).

In standalone positioning, the receiver receives the transmission on position and time of the satellite, calculates the time for the signal to travel from the satellite to the receiver, and converts the information into distance. The receiver determines its position based on its distances from four or more satellites. The calculated position has an error of about 10 m because of satellite positional errors, and radio wave delays when the satellites pass through the troposphere or ionosphere.

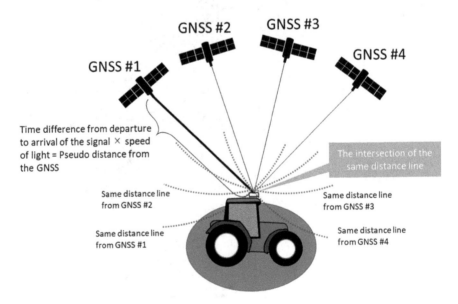

Fig. 6.5 Principle of satellite positioning system

In relative positioning, two or more receivers simultaneously track four or more satellites to determine their relative positions. This method is more accurate than standalone positioning. Relative positioning can use differential GNSS (DGNSS) to obtain the relative position by performing standalone positioning with respect to multiple receivers. The position is determined with an error of a few meters. Additionally, interference positioning is also one of the relative positioning. The interference positioning obtains the difference in distance (path difference) between multiple receivers and the satellite; the relative position can be determined based on the phase of the carrier wave. The real-time kinematic GNSS (RTK-GNSS) is an example of position determination based on the satellite data and reference station data in real time with an error of a few centimeters.

6.2.1.2 Improvement of Positioning Accuracy

The positioning accuracy by using the satellite positioning system has been improved year after year. The US GPS began to operate officially in December 1993; however, the precision error by using the single positioning method had been 100 m in horizontal distance until the 1990s when the military precision degradation measures known as SA (selective availability) became available. The current precision error is several tens of meters after the countermeasure was abolished. In addition, the number of available satellites as well as the numbers and frequencies of signal has increased due to the use of multi-GNSS aforementioned (GNSS Technologies 2016). As a result, the accuracy, reliability, and stability of the GNSS are greatly improved.

The QZSS of Japan was started by launching the first quasi-zenith satellite "Michibiki" in 2010. The positioning accuracy of several centimeters is made possible by using the satellite-based augmentation system (SBAS) to augment signals from the satellite (Sakai 2016). Quasi-zenith satellites stay long in the vicinity of Japan; they may be visibly in the range of ascending vertical angles from 60 ° to 90 ° so that the satellite signals can be received even in the canyon of high buildings and deep forests. L1C, L2, and L5 transmit signals that are the same as those transmitted from GPS and another reinforcement signal with L6 (LEX) of 1278.75 MHz that has the same frequency as Galileo enhance high positioning accuracy (Table 6.1).

Table 6.1 Signals of satellite positioning systems

	Frequency (MHz)			
	1176.45	1227.60	1278.75	1575.42
GPS	L5	L2C		L1C/A
GZSS	L5	L2C	L6(LEX)	L1C
Galileo	E5a, E5b		E6	E1

The standalone positioning method needs to receive the signals from four satellites at least in principle. Code of the same length involving the information on the date and clock time, tracking positions (Ephemeris) of the principal satellite, and tracking positions (Almanac) of all other satellites are repeatedly broadcasted from the satellite transmitting devices at the same wavelength such as L1 of 1575.42 MHz. The pseudodistance from the satellite to the receiver is obtained using the time delays (difference) between the time of signal leaving the satellite and reaching the receiver on the ground. Solving the unknown three-dimensional coordinates and the clock error of the receiver by using the following simultaneous Eq. (6.1), the intersection of the same distance lines is found.

$$r_1 = \sqrt{\left(x_1 - x\right)^2 + \left(y_1 - y\right)^2 + \left(z_1 - z\right)^2} + s$$

$$r_2 = \sqrt{\left(x_2 - x\right)^2 + \left(y_2 - y\right)^2 + \left(z_2 - z\right)^2} + s$$

$$r_3 = \sqrt{\left(x_3 - x\right)^2 + \left(y_3 - y\right)^2 + \left(z_3 - z\right)^2} + s$$

$$r_4 = \sqrt{\left(x_4 - x\right)^2 + \left(y_4 - y\right)^2 + \left(z_4 - z\right)^2} + s \tag{6.1}$$

Where

r_i: pseudodistance from satellite i to the receiver
x_i, y_i, z_i: x, y, z position of satellite i
x, y, z: x, y z position of the receiver
s: clock error of the receiver

One of the reasons that the standalone positioning method has a positioning error of several tens of meters is due to the clock error described above. Further, the tracking position information of the satellite in the navigation code message is another source of error factor because the ephemeris parameter is updated once every 2 h, and the oral history almanac data is updated once every 6 days.

6.2.2 Farming Using Driving Support System Controlled by GNSS

6.2.2.1 Driving Support System

A driving support system (DSS) incorporates a farming guidance system (FGS) based on GNSS and automatic driving assistance (ADA). Using the DSS, even an inexperienced operator can operate a tractor as efficiently as a skilled operator. DSSs have been widely adopted to overcome labor shortages. Figure 6.6 summarizes where the DSS offers benefits on crop rotation with improved accuracy of operating and tracking the machinery in a paddy field.

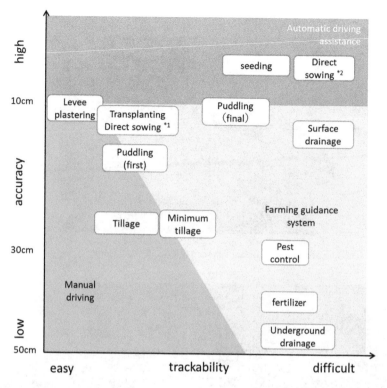

Fig. 6.6 The region in which a driving support system offers benefits on crop rotation in a paddy field. *1: Direct sowing of rice in flooded paddy field. *2: Direct sowing of rice in a well drained paddy field

6.2.2.2   Farming Guidance System

Many of the FGSs currently available on the market use DGNSS. An FGS consists of a GNSS antenna and a monitor to display work routes and trajectories (Fig. 6.7). The FGS plots the work routes and trajectories using the GNSS position information while the operator supervises the work. The operator first lines up the equipment with the field edge without guidance. The implied orientation becomes the reference line, and equally spaced work routes parallel with the reference line are created and displayed on the monitor. By following these work routes, the machine can perform its operations at equal intervals. The author and colleagues tested a spreading material from a broadcaster. Without the FGS, the material was spread unevenly, whereas with FGS, the material was spread more evenly (Fig. 6.8). In addition, reducing the scattering unevenness without using the FGS makes it necessary to run the machine in orthogonal directions. However, by using the FGS, it is possible for the machine to follow a single path thus reducing the working time by 40%, improving work efficiency, reducing costs, as well as functioning as tillage, puddling, and pest control.

Fig. 6.7 Example of a farming guidance system (GPAS, Kubota Corporation)

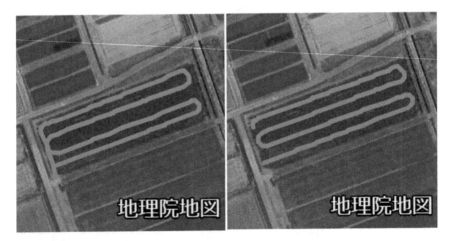

Fig. 6.8 Left: Manually guided work route. Right: Route by using farming guidance system

6.2.2.3 Automatic Driving Assistance

The GNSS-based automatic steering ADA consists mainly of a steering control, a GNSS antenna, a monitor, and a controller, which includes a gyroscope and an accelerometer (Fig. 6.9). Establishing the reference line is discussed in above sections. The ADA steers with an electric motor guided by GNSS thus allowing the operator to concentrate on other tasks while driving the machine to improve the work quality and volume. By using the RTK-GNSS control, the machine can drive autonomously with an accuracy of 5 cm. Table 6.2 compares the accuracy of the sowing work by a skilled operator and an unskilled operator using an ADA. The results show that both skilled and unskilled operators achieve the same accuracy.

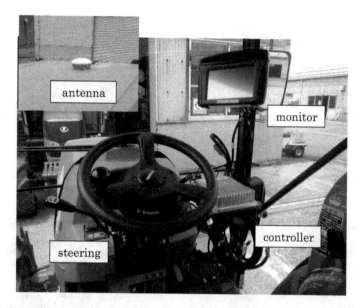

Fig. 6.9 Example of automatic driving assistance (CFX750 and Autopilot Motor Drive system, Nikon-Trimble Co., Ltd.)

Table 6.2 Comparison of sowing accuracy by skilled operator and unskilled operator using automatic driving assistance

Operator	Drive	Work area (m²)	Work efficiency (%)	Work accuracy (%)
Skilled	Auto	3910	85.9	93
Unskilled	Manual	2460	64.8	27
	Auto	2580	67.2	92

Although the ADA system is highly effective for farming work, it is not popularly used in medium-scale and distributed farmland in southern Honshu (main island of Japan) because the system is costly. To improve the cost-effectiveness of this system, the writer and colleagues examined the practice of sharing one ADA among a 37 kW tractor, a 56 kW tractor, and a multipurpose rice transplanter. Switching between machines was easy; it took 24 min to move the ADA from one machine to another and about 2 min to reset the guidance monitor (Fig. 6.10). In a test field at the Hokuriku Research Station, Joetsu city, the writer and colleagues used the ADA for tillage (37 kW tractor), puddling (56 kW tractor), and transplantation or direct sowing of rice with multipurpose rice transplanter. Sharing the ADA among the machines allowed us to apply the same reference line and start work route for all operations from the correct position each time. The straight operation made all tillage and puddling uniform, thus allowing seeds to be sown or seedlings to be transplanted at uniform intervals. During the automatic operation, the operator was able to pay attention to other tasks such as replenishing seedlings. The ADA

Fig. 6.10 Sharing the automatic driving assistance between multipurpose rice transplanters (top) and tractors (middle, 37 kW; bottom, 56 kW)

thus enhanced efficiency not only by improving the work performance but also by freeing up the operator to do other important tasks at the same time.

6.2.2.4   Use of GNSS in Agriculture

Since its introduction in 1985, GNSS has been used in agriculture such as the "precision farming" in the United States and Europe during the early 1990s (Noguchi 1996). Precision farming relies on the collected field information to support work management and the use of GNSS to obtain accurate real-time positions in a field; both GNSS and field information are indispensable to precision farming.

GNSS is also crucial for the realization of farming robots, which are currently under development. Research on autonomous vehicles such as robots in Japan began officially in around 1985 and developed along with advances in GNSS positioning technology. The technical aspects of the autonomous operation of farm machinery such as tractors, transplanters, and combine harvesters, using GNSS positioning are in place. Because such robots use high-precision GNSS and attitude sensors, they may reach faster speeds if an error within 5 cm is allowed.

GNSS is also indispensable for small unmanned aerial vehicles (drones). Drones equipped with a compass and an altimeter can fly autonomously under the control of GNSS. Advances in miniaturization and cost reduction now allow drones to be introduced into agriculture for precise remote sensing such as discrimination of crops and weeds.

The application of GNSS for various uses in agriculture and its continuing improvement make GNSS an important technology for using agricultural machinery in farming work.

6.3 Advanced Fruit Production Using Biological Monitoring System

6.3.1 Feature of Fruit Production Around a City

Producing fruits around a city is popular in Japan. The advantage of such a practice is that a large consuming community is within a short distance of the orchard. The demand of fresh fruits that can be eaten directly without cooking is high especially because fully ripened fruits are difficult to transport over long distances to reach consumers living in a city.

According to the information provided by Tokyo Metropolitan Government (2018), fruits that are more likely cultivated in Tokyo are chestnuts, persimmons, and blueberries in that order. In recent years, the cultivation area of blueberries has been rapidly increasing (Tokyo Metropolitan Central Wholesale Market 2017, Fig. 6.11). The cultivation area of blueberries in Japan was 521 ha in 2003; it doubled to 1133 ha in 2013 with the annual yield also continues to increase for more than 30 years (Ministry of Agriculture, Forestry and Fisheries of Japan 2015). Increasing the cultivation of blueberries in the urban and its suburban regions also supports the increase of agricultural production. In addition, blueberry fields have attracted attention as tourism orchards, and the fruits can be either consumed directly on-site or processed in cake shops or bakeries. The antioxidant effect of blueberry makes it a healthy fruit so that in addition to its high economic value, blueberry is quite popular among health-conscientious consumers. The unit price of

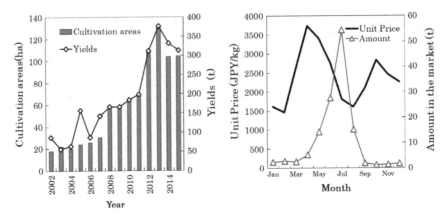

Fig. 6.11 Cultivation area and yields during the period between 2002 and 2015; Monthly variations of blueberry amount and unit price in Tokyo (2017)

blueberries in Tokyo market was 3740 yen/kg in April and even higher during off season (Tokyo Metropolitan Central Wholesale Market 2017). Blueberries can be considered as a representative fruit being cultivated in a city.

However, the fruit production around a city has some issues that need to be addressed. When growing agricultural products in the city suburban, one should take some precautions. For examples, when the farmers spray pesticides and apply fertilizers in the field that is close to houses, the residents, who are not farmer in and around the city, will complain about the operations. Hence, the operations need to be carried out in a closed or semi-closed cultivation system using the advanced greenhouse technology. Normally, the initial and operating costs of an advanced greenhouse are high; the high costs can be paid for by growing high-priced fruits such as blueberries. Precise management to monitor biological information on potted fruit trees is expected to improve fruit qualities and reduce the production costs. Greenhouses based on advanced technology are expected to promote tourism-related agriculture in addition to producing food-related fruits. To improve the functionality and annual production of blueberries, developing technologies to monitor plant growth and optimize cultivation environment and improving fruit varieties and cultivation techniques are of great importance. Appropriate management of pot cultivation to grow fruit trees in a greenhouse can achieve advanced control of the plant growth environment. The shallow-rooted fibrous roots of blueberries are sensitive to drying and high temperature so that precise cultivation management of the plant is required.

6.3.2 Appropriate Environmental Control in Growing Fruits

A greenhouse with advanced environmental control, which can be considered as a factory for fruits production, has attracted increasing attention because of its capability to produce fruits with high quality in consistent quantity year-round. An optimum environmental control in the greenhouse results in improved qualities of agricultural products at reduced cultivation cost.

To conserve resources for plant cultivation, the plant growth and resource consumption in the cultivation environment need to be monitored. Some researchers believe that plants themselves may offer answers. Of course, it is important to know the environment of cultivation, for example, the air temperature and how much water is used to plants per day. The information of the light environment and the daily water demand for plant growth is important for practicing precise management. However, the most important information is the state of the cultivated plants. Normally, professional farmers know how to cultivate the plant based on their years of experience. They will adjust the controlling parameters such as the shading curtain and the irrigation apparatus according to the needs of the plant to grow. For inexperienced and novice farmers, the information collected by observing the plant growth will be quite useful. If the information can be collected automatically and used for implementing agricultural management, fruit cultivation will take a further step toward innovation of fruit production (Fig. 6.12). More specifically, farmers can use the information appropriately to save labors and add values to their agricultural products.

6.3.3 Data Collection Methods of Biological Information

In concrete terms, one has to ask what kind of biological information from plant is necessary and which parts of plants will provide useful information. The various information from a plant body such as leaves, branches, roots, fruits, and flowers

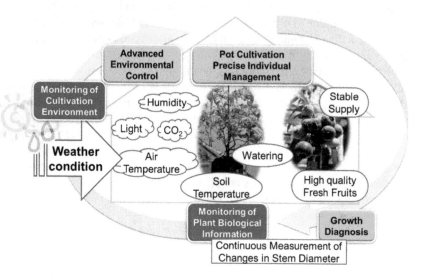

Fig. 6.12 Relationships between plant biological information and cultivation environment in a greenhouse

may be collected. The information includes photosynthetic capacity, amount of transpiration, dynamics of water movement inside a plant body, and chemical components of products, among the many others.

The water balance is an important factor that relates to restriction of photosynthesis or growth disturbance, and various biological information concerns the water movement in an individual plant. According to Muramatsu (2007), a number of methods are available for measuring the water state in a plant. These methods include impedance method, time-domain reflectometry (TDR) method, microwave method, AE method, fluorescence method, porometer method, pressure chamber method, heat pulse method, stem heat balance method, leaf temperature measurement method, and near-infrared spectroscopy method. Real-time monitoring of the water movement in a plant body will avoid disturbance to the plant growth caused by a sudden stress so that a stable production is warranted. In some fruit tree and vegetable cultivations, consistently high quality of fruit production can be achieved by monitoring the cultivation environment and the water state inside the plant. Many researchers have reported the various advantages and influences of irrigation control on plant growth, flower development, and yields of blueberries (Spiers 1996 and 1998, Haman et al. 1997).

According to Fujita et al. (2011), water stress in trees can be accurately estimated using the time-domain reflectometry (TDR) method in which the soil and tree moisture contents are measured to provide practical diagnostic techniques for supporting the orange cultivation. Oda et al. (1993) indicated that measuring chlorophyll fluorescence may be a practical method of detecting heat stress or injury to the photosynthetic organs. Also, the results reported by Nakahara and Inoue (1997) suggested that the management of water stress in tomato cultivation could be one of the most feasible applications of infrared thermometry. However, implementing these measurement methods may easily damage or destroy a plant by installing sensors on the plant in addition to difficulties for carrying out continuous measurement.

The stem diameter method, which hardly causes serious damages to the plants by using simple devices, is used to check changes in plant stem diameter. The stem of fruit trees expands during normal growth but shrinks sensitively in response to adverse environmental factors. Some techniques have been developed to detect changes in plant stem diameter and use the results as indicators of the required irrigation (Goldhamer and Fereres 2001, Nortes et al. 2005). Advanced controls based on the growth information of individual plants are more effective in enhancing plant growth and product quality. If continuous measurement of stem diameter is possible for individual plants, the information can be used for cultivation management on the precise growth of plant, water state of stem, and diagnosis of growth. An example of controlling the environment in a greenhouse based on plant biological information is shown in the following section.

In open field, the cultivation environment around the plant is significantly influenced by the weather condition. When weather changes, the ambient conditions around the plant such as solar insolation, air temperature, relative humidity, and concentration of CO_2 are also changed. Furthermore, changes in soil moisture and soil temperature in the cultivation pot may occur. Plants can respond to such environmental changes; then finally fruits can be grown through physiological reactions including transpiration and photosynthesis. In a greenhouse, continuous monitoring of the cultivation environment and implementing advanced environmental control enable the farmer to practice precise control of the environment around the pot so that accurate management of individual plants becomes possible. By creating a suitable environment for individual plants, high-quality fruits can be produced in stable quantity. In order to determine the suitable environment for an individual plant, monitoring the plant growth continuously to collect the plant biological information is of great importance.

As the plant biological information is concerned, the writer and colleagues focused on continuously monitoring changes in stem diameter of blueberry. The stem diameter may shrink during daytime and recover at night, and this minimal change responds sensitively to changes in the cultivation environment. Because the displacement is in micron unit, it cannot be observed visually by the naked eye. By analyzing such invisible changes in stem diameter with respect to the environmental factors around the plant, one can diagnose the optimum environment for blueberries cultivation. Moreover, predicting the change in stem diameter based on the monitoring results allows one to perform cultivation control in advance to avoid inhibitions to the plant growth.

6.3.4   Monitoring Experiment of Stem Diameter Variations

6.3.4.1   Importance of Continuous Monitoring

Changes in stem diameter based on growth monitoring is used as an effective indicator of individual fruit trees (Kikuchi et al. 2014a, b and 2017). The writer and colleagues implemented this method in actual cultivation of blueberry plants. Stem diameter is generally known as one of the most sensitive indices of water stress; it is a useful indicator to assist in practicing irrigation management in fruit production (Goldhamer and Fereres 2001, Nortes et al. 2005). Diurnal diameter changes measured at the phloem and xylem surfaces have a similar pattern as diurnal changes in leaf water potential (Ueda and Shibata 2001). Simonneau et al. (1993) reported that the trunk diameter varies with internal water balance of the shoot because changes in stem diameter are caused by plant physiological effects in addition to environmental conditions.

There are many types of sensors, such as strain gage (Ueda and Shibata 2002), dendrometer (Miralles-Crespo et al. 2010), linear variable differential transformer (LVDT) (Goldhamer and Fereres 2001), laser (Oishi 2002), and load cell (Iwao and Takano 1988) that can be used for monitoring the stem diameter. Considering the long-term measurements for cultivation managements, the monitoring results can be used in the control of cultivation environments such as in greenhouses where an inexpensive and a highly accurate measurement are required. One measurement method uses a strain gauge that is attached directly to the plant body. This method is advantageous because its operation is simple and inexpensive. However, in order to measure the stem diameter precisely and continuously, it is important not to damage the bark of the trunk and eliminate the influence by external factors such as temperature and humidity. Long-term measurements for more than 1 month are practical in a greenhouse where the monitoring sensor is not influenced by changes of the solar radiation and ambient temperature.

The dendrometer attached to a tree by using a tape or a wire winding around the tree trunk is usually for evaluating the growth of large trees. There is a report that dendrometers can be used to measure ornamental shrubs of 1–2 cm in diameter for potted cultivation (Miralles-Crespo et al. 2010). Whether the dendrometer can be used to measure small fruit shrubs such as blueberries of less than 1 cm in diameter with many branches is doubtful. Some dendrometer measurements also use strain gauges and the LVDT sensor. A laser displacement sensor that can measure the stem diameter nondestructively with high accuracy without contacting the plant is very expensive. Moreover, this method is susceptible to peripheral factors such as solar radiation and dusts among others for doing continuous measurement even in a greenhouse. The sensor needs supporting pillars to be mounted on the plant. Further, the measurement range of specific sensors is quite narrow.

A continuous measurement is made possible by using a strain gage sensor with higher accuracy but relatively lower cost than a laser sensor. This method has advantages in that it is less influenced by the ambient temperature and easy to be attached to arbitrary stems (Kawai et al. 2003). Thus, in this study, the measurement method using a strain gauge along with a cantilever-type displacement sensor was adopted for measuring changes in stem diameter.

6.3.4.2 Monitoring Method of Stem Diameter

A small load cell was examined in this experiment to measure the real-time stem diameter displacement with high precision. Rabbit eye blueberry plants (*Vaccinium virgatum* 'Homebell') planted in pots of 280 mm diameter filled with mixture of peat moss and Kanuma soil were selected as the target plants. The blueberry was grown in a greenhouse using automatic irrigation for eight times a day.

The device for measuring stem diameter was installed on the main stem and a branch of a selected blueberry plant to carry out the long-term measurement. The diameter of the main stem near the root was 8.9 mm and the diameters of two selected branches were 5–6 mm. The branch was a younger part that extended from the main stem this year.

Microscopic diurnal changes in stem diameters can be continuously and nondestructively monitored by using the load cell (ULA-10GR, Minebea) developed by Iwao and Takano (1988). Changes in stem diameters were detected by the displacement from the original position of cantilever; the information is then amplified (TEAC, TD-510), converted with an A/D conversion, and finally recorded in a data logger (memory high logger 8430, HIOKI). This system can detect imperceptible changes in blueberry stem diameter (Kikuchi et al. 2017). The load cell was bonded to the blueberry plant between the stem and the cantilever of the load cell by using rubber bands and a piece of silicone tube clipped with a pinchcock. The load cell moves with changes in stem diameter, while its cantilever remains steady so that it detects the relative position difference between the stem and the cantilever as the plant stem diameter changes.

Figure 6.13 shows typical changes in stem diameter. Some indices were determined as the biological information with respect to changes in stem diameter. Maximum daily stem shrinkage (*MDS*) is the daily differences between the maximum and the minimum daily stem diameters that is defined as

$$MSD = MXSD - MNSD.$$

Where

MXDS is maximum daily stem diameter
MNDS is minimum daily stem diameter

The change rate of stem diameter (*CRD*) is defined as the diameter change per unit time. The relationship between stem shrinkage and water potential can be expressed by linear regression lines (Imai et al. 1990). The diurnal changes in diameter match the diurnal changes in water balance of the sap flow and transpiration (Ueda and Shibata 2001). Some researchers proposed irrigation scheduling protocols based on parameters including maximum shrinkage and growth rate, among others. (Goldhamer and Fereres 2001, Nortes et al. 2005).

6.3.4.3 Hourly Changes in Stem Diameter

The results of changes in stem diameter were strongly influenced by the sun radiation (Fig. 6.14). The diurnal stem variation shows that the stem diameter starts to decrease immediately after sunrise; it continues to shrink sharply to reach the

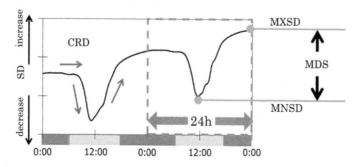

Fig. 6.13 Definition of indices on changes in stem diameter. *MDS* maximum daily stem shrinkage, *MXDS* maximum daily stem diameter, *MNDS* minimum daily stem diameter, *CRD* change rate of stem diameter per unit time

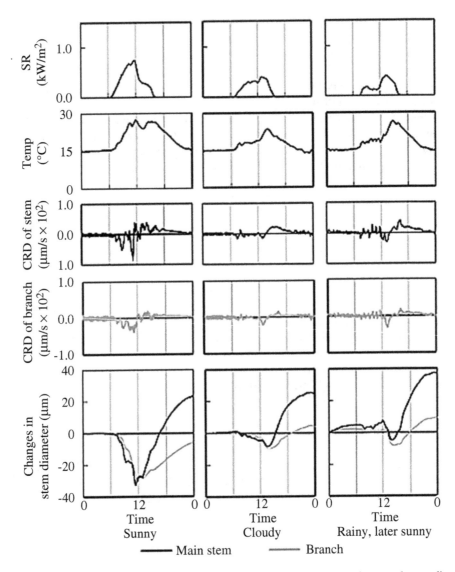

Fig. 6.14 Changes in stem diameter and environmental conditions under various weather conditions. *SR* solar radiation, *CRD* change rate of stem diameter per unit time

minimum level at noon. From noon to sunset, it increases to the original diameter before the shrinkage occurred. It then continues to increase slightly from sunset to midnight. (Kikuchi et al. 2012). The diurnal changes in the branch from midnight to noon shows the same tendency as the diameter changes in the main stem. The increment of the diameter changes in the branch from noon to sunset is smaller than the changes in the main stem.

The results obtained in this study are consistent with the findings of various previous analyses on the relationship between water status and diurnal changes in stem diameter. The difference of the diameter changes between the main and the branch stems depends on the number of leaves and the leaf areas from where the instrument is attached to the top of the plant. Imai et al. (1990) has pointed out that the change in stem diameter is highly correlated with the leaf water potential. Ueda and Shibata (2002) also reported a close relationship between changes in stem diameter and the water content of the plant body. One could monitor growth changes in the main stem with higher sensitivity than measuring the branch diameter. More significant changes may be obtained by measuring the diameter change at the main stem for a long-term monitoring of the blueberry plant biological information; however, the measurement at branch may reveal other relevant information. Blueberry plants begin to fruit at the tip of the shoot where the wilting and the tip burning are likely to occur. The long-term measurement reveals that diurnal changes in the diameters can be clearly observed at both the main stem and the branch, but the branch has slower diameter growth than the main stem. In this study, the long-term measurement of the stem diameter was observed successfully under the environmental conditions in actual cultivation.

The characteristic environmental factors were extracted from diurnal changes in stem diameter. On cloudy days, the difference between the changing rates in stem diameters for the main stem and the branch was relatively small, whereas the difference became more obvious on sunny days. When the sunshine continues for days, the branch diameter decreases without recovering to the original diameter before the shrinkage occurred. This observation may be caused by a water shortage near the tip of the stem due to heavy evapotranspiration from the blueberry plant. These results show that the stem diameter change can be used as an indicator for diagnosing the growth of blueberry plant.

6.3.5 Future of Potted Fruit Production

Various inexpensive and easy to operate sensors for monitoring agricultural environments have been available on the market. Microcomputers used for collecting and processing data have also become inexpensive. Under such circumstances, the importance of data collection and the real-time analysis will be evaluated. Implementing appropriate control of the cultivation environment based on biological information of each plant is no longer a dream. Figure 6.15 shows the schematic diagram of an intelligent pot with advanced controlling technology in a greenhouse. In the future, a set of sensors attached to the plant and the soil in the pot provide the relevant information automatically. An enormous quantity of real-time data will be analyzed by using a PC, and the cultivation environment will be controlled based on the data thus collected and processed. Using artificial intelligence (AI), adequate management strategies for managing the cultivation can be provided to farmers who are not familiar with agricultural operations. With an improved PC processing capa-

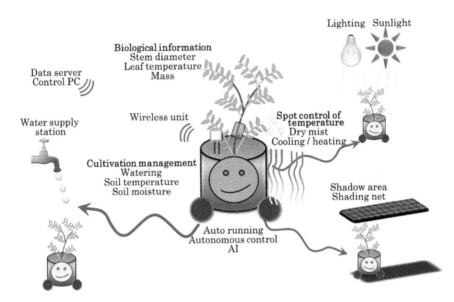

Fig. 6.15 Concept of an intelligent pot with advanced controlling technology for fruit production in a greenhouse

bility, biological images of the plants analyzed in real time will be visually displayed on a monitor in the near future.

The movable intelligent pot for fruit cultivation is an idea that automatically provides information obtained with multiple sensors to formulate control of the cultivation environment. For example, when feeling thirsty, one will go to the water source to drink; an intelligence pot behalves similarly. It senses the lack of water in fruit plant and moves to the watering station to receive the necessary amount of water. Of course, the pot is capable of predicting the water need in advance to keep the plant from suffering heatstroke. According to changes in temperatures of the cultivation environment, the spot control system such as spot cooler and heater is expected. In response to changes in solar radiation when a cultivation pot is exposed to direct sunlight in a greenhouse, the biological plant information may be fed back to the control system to adjust shading curtain for maintaining the greenhouse temperature. The wireless communication system enables autonomous control of individual plant based on the biological information collected and analyzed. By setting the control indices at certain predetermined levels, farmers can optimally manage the cultivation pot without depending on advanced knowledge. For example, management settings such as applying appropriate fertilizer, as well as suitable time for light blocking, watering, and other tasks, will be considered as the control indices. Like human beings who need balanced nutrient to grow healthily, plants need balanced fertilizer to grow healthily as well. Thus, when an irregularity is detected in the diurnal changes of the stem diameter, the nutritional state of or pest damages to the plant needs to be diagnosed. Appropriate prescriptions for applying fertilizer or controlling pest will be implemented for individual plant. In the future, such advanced control of the cultivation system can be established for greenhouse production of both organic products and ordinary horticultural products. Even in current greenhouses, control of growth environment can be made based on the information of plants growing in the cultivation system. In the future, establishing a more stable system for all year production with high quality by implementing an autonomous control of local cultivation environments based on biological plant information with an AI system is required for saving production energy and maximizing cost-effectiveness.

In fact, currently very few organic agricultural products are grown officially in greenhouses. From now on, engineering approaches should focus on carrying out organic horticulture in greenhouses. A well-designed greenhouse enables the production of organic crop by employing high technologies to operate the greenhouse with less CO_2 emission based on not only environment conditions but also biological plant information. Such ideas, technologies, and systems need to be considered by industries to produce organic food cost-effectively in greenhouse. The knowledge that combines agricultural production and engineering practice is of great importance to achieve more cost-effective production of high-quality organic food in greenhouse.

References

Blackmore S (2016a) Robotic agriculture: smarter machines using minimum energy, presented on April 21, 2016. https://www.youtube.com/watch?v=4RiWMOz3J4w. Accessed 3 Dec 2018

Blackmore S (2016b) Farming with robots. https://www.bing.com/videos/search?q=simon+blackmore&view=detail&mid=6239F14A42887F729E066239F14A42887F729E06&FORM=VIRE. Accessed 3 Dec 2018

Chosa T (2017) Why do we try to do agriculture with a small robot? In: Proceedings of the 76th annual meeting of The Japanese Society of Agricultural Machinery and Food Engineers 193. (Presentation No. 7–8) (in Japanese)

Chosa T (2018) Lets's start small farming? In: Proceedings of the 21st Techno-Festa, Symposium of the Japanese Society of Agricultural Machinery and Food, (in press). (in Japanese).

Chosa T, Furuhata M, Omine M, Matsumura O (2009) Development of air-assisted strip seeding for direct seeding in flooded paddy fields seeding machine and effect of air assistance. Jpn J Farm Work Res 44(4):211–218. in Japanese with English summary

Chosa T, Tojo S, Rojas J L P (2015) An attempt of the air-assisted strip seeding for direct seeding using a knapsack power applicator, Proceedings of Joint Conference on Environmental Engineering in Agriculture 2015 (on CD-ROM). (Presentation No. B204)

Faulkner EH (1943) Plowman's folly, the University of Oklahoma Press, Publishing Division of the University. Manufactured in the USA

Fujimori K, Chosa T, Tojo S (2017a) Evaluation by simulated seeding distribution envisioning random traveling of small field robot. Proceedings of the 53rd annual meeting of Kanto branch, The Japanese Society of Agricultural Machinery and Food Engineers, 32–33. (Presentation No. A13) (in Japanese)

Fujimori K, Chosa T, Tojo S (2017b) Seeding simulation by small field robot. In: Proceedings of the 76th annual meeting of The Japanese Society of Agricultural Machinery and Food Engineers, 186. (Presentation No. 7–1) (in Japanese)

Fujita A, Nakamura M, Kameoka T (2011) Soil moisture measurement to support production of high-quality oranges for information and communication technology (ICT) application in production orchards. Agric Inf Res 20:86–94. https://doi.org/10.3173/air.20.86

GNSS Technologies Inc. (2016) Fundamental knowledge of GNSS (in Japanese). https://www.gnss.co.jp/?page_id=5341. Accessed 10 Feb 2019

Goldhamer DA, Fereres E (2001) Irrigation scheduling protocols using continuously recorded trunk diameter measurements. Irrig Sci 20:115–125. https://doi.org/10.1007/s002710000034

Haman DZ, Smajstrla AG, Pritchard RT, Lyrene PM (1997) Response of young blueberry plants to irrigation in Florida. HortScience 32:1194–1196

Imai S, Iwao K, Fujiwara T (1990) Measurements of plant physiological information of vine tree and indexation of soil moisture control (1). Environ Control Biol 28:103–108

Iwao K, Takano T (1988) Studies on measurements of plant physiological informations and their agricultural applications (1). Environ Control Biol 26:139–145

Kato T, Chosa T, Tojo S, Yoshikawa M, Mochimaru T (2016) Influence of mowing using robot lawn mower on vegetation in weeds. Japn J Farm Work Res 51(1):23–24

Kawai K, Ito J, Ohkura K, Fujita K (2003) Measurement of changes in stem diameter under different temperature by an unbonded type gauge device. Environ Control Biol 41:289–294

Kikuchi R, Chosa T, Tojo S (2012) Root temperature influences growth of blueberry stems. CIGR-AgEng2012, International conference of agricultural engineering, Val Spain, July

Kikuchi R, Chosa T, Tojo S (2014a) Relation between changes in stem diameter of blueberry and water balance of absorptions and evapotranspiration. International conference on agricultural engineering, AgEng 2014, Zurich, 6–10 July

Kikuchi R, Chosa T, Tojo S (2014b) Development of continuous water supply system for measurements of water balance of absorptions and evapotranspiration of blueberries. International conference on agricultural engineering, AgEng 2014, Zurich, 6–10 July

Kikuchi R, Chosa T, Tojo S (2017) Continuous measurement for stem diameter of blueberry plant with small load cell. J Jpn Soc Agric Mach Food Eng 79:365–373

Komatsuzaki M (2007) Ecological significance of cover crop and no tillage practices for ensuring sustainablity of agriculture and eco-system service. In: Chen J, Guo C (eds) Ecosystem ecology research trends. Nova Science Publishers, New York, pp 177–207

Kuroda H (2017) Verification of power supply to lawn mowing robot by independent power supply (solar + storage battery). In: Proceeding of the 76th annual meeting of the Japanese Society of Agricultural Machinery and Food Engineers, 187. (No. 7–2) (in Japanese)

MAFF (2015) Ministry of Agriculture, Forestry and Fisheries of Japan, 2002–2015. The statistics of production dynamics of locally produced fruit tree. http://www.maff.go.jp/j/tokei/kouhyou/tokusan_kazyu/

Miralles-Crespo J, Sanchez-Blanco MJ, AN G et al (2010) Comparison of stem diameter variations in three small ornamental shrubs under water stress. HortScience 45:1681–1689

Muramatsu N (2007) Recent measurement of moisture content in body of fruit tree. Agric Hortic 82:947–955. (in Japanese)

Nagasaka Y, Taniwaki K, Otani R, Shigeta K (1999) The development of autonomous rice transplanter (Part 1) – The location of the rice transplanter by a real-time kinematic GPS –. Journal of Japanese Society of Agricultural Machinery 61(6):179–186. (in Japanese with English abstract

Nakahara M, Inoue Y (1997) Detecting water stress in differentially-irrigated tomato plants with infrared thermometry for cultivation of high-Brix fruits. J Agric Meteorol 53:191–199. https://doi.org/10.2480/agrmet.53.191

Noguchi N (1996) An over view of the global positioning system (GPS) and application for agriculture – Advanced technology in the near future. J Jpn Soc Agric Mach 58(4):130–134

Nortes PA, Perez-Pastor A, Egea G et al (2005) Comparison of changes in stem diameter and water potential values for detecting water stress in young almond trees. Agric Water Manag 77:296–307. https://doi.org/10.1016/j.agwat.2004.09.034

Oda M, Li Z, Tsuji K et al (1993) Effects of humidity and soil moisture content on chlorophyll fluorescence of cucumber seedlings exposed to high air temperature. J Japn Soc Hortic Sci 62:399–405. https://doi.org/10.2503/jjshs.62.399

Oishi N (2002) Development of irrigation control system in response to plant water stress in tomato hydroponics (1). Environ Control Biol 40:81–89

Pedersen SM, Fountas S, Have H, Blackmore BS (2006) Agricultural robots—system analysis and economic feasibility. Precis Agric 7(4):295–308

Saito M, Tamaki K, Nishiwaki K, Nagasaka Y (2012) Development of an autonomous rice combined harvester using CAN bus network. J Japn Soc Agricult Mach 74(4):312–317. (in Japanese with English abstract

Sakai T (2016) Status of the Japanese QZSS Program. Munich Satellite Navigation Summit, Munich, Germany, March 2016. https://www.enri.go.jp/~sakai/pub.htm. Accessed 10 Feb 2019

Simonneau T, Habib R, Goutouly J-P, Huguet J-G (1993) Diurnal changes in stem diameter depend upon variations in water content: Direct evidence in peach trees. J Exp Bot 44:615–621. https://doi.org/10.1093/jxb/44.3.615

Spiers JM (1996) Established "Tifblue" rabbiteye blueberries respond to irrigation and fertilization. HortScience 31:1167–1168

Spiers JM (1998) Establishment and early growth and yield of "Gulfcoast" southern highbush blueberry. HortScience 33:1138–1140

Tokyo Metropolitan Central Wholesale Market (2017). Market statistics

Tokyo Metropolitan Government, Bureau of Industrial and Labor Affairs (2018) Fruit tree agriculture promotion plan of Tokyo

Ueda M, Shibata E (2001) Diurnal changes in branch diameter as indicator of water status of Hinoki cypress Chamaecyparis obtusa. Trees - Struct Funct 15:315–318. https://doi.org/10.1007/s004680100113

Ueda M, Shibata E (2002) Water status of the trees estimated from diurnal changes in stem and branch diameters using strain gauges. J tree Heal 6:75–84

Yukumoto O, Matsuo Y, Noguchi N, Suzuki M (1998) Development of tilling robot (Part 1) – concept of control algorithm and task planning using position sensing system and geomagnetic direction sensor. J Jpn Soc Agric Mach 60(3):37–44. (in Japanese with English abstract

Chapter 7
Soil Health and Carbon Sequestration in Urban Farmland

Toru Nakajima

Abstract Changes of the urban land use could lead to significant impact on the ecosystem service because of losses of habitat, biomass, carbon pool in the soils, and biodiversity. Depletion of the carbon pool in the urban area could amount to 0.05 petagrams (Pg) C per year resulting in a total loss of 5% of the total emissions from tropical deforestation. The soil fertility of farmland is very important to maintain sustainable qualities of agricultural products. Carbon sequestration in soil is one of several alternatives to increase the soil fertility and diversity of farmlands in the urban area.

The idea of soil health/quality is related to the capacity of soils to function properly in support of the important ecosystem services needed to sustain soil productivity and maintain environmental quality. The soil health assessment will enable us to understand future trends in soil physical, chemical, and biological properties as well as its functions overtime, to evaluate whether soil health under different agriculture managements is aggrading, sustaining, or degrading. The soil physical properties directly affect the plant root development, water availability, and aeration that in turn influence the plant productivity in agricultural land. Soil chemical properties influence nutrient availability, salinity, and alkalinity which directly affect plant/crop productivity and quality. The soil organic matter impacts soil chemical properties and has profound effects on plant growth and yield. Soil biological properties and processes influence both soil physical and chemical properties and directly affect the plant health and repeated cultivation damage. Soil microorganisms play an important role to break down organic matters contained in the soil. Applying compost, cover crop, and biochar will increase soil microbial activities and mediate diseases caused by soil-borne pathogens in field trials. The soil biological properties are crucial to soil health, plant health, and crop productivity in organic urban farming.

The assessment of carbon stocks in an urban soil is more complicated than assessing the general agriculture soils. The carbon sequestration rate is determined

T. Nakajima (✉)
Department of Bioproduction and Environment Engineering, Tokyo University of Agriculture, Setagaya-ku, Tokyo, Japan
e-mail: tn206473@nodai.ac.jp

© Springer Nature Singapore Pte Ltd. 2020
S. Tojo (ed.), *Recycle Based Organic Agriculture in a City*,
https://doi.org/10.1007/978-981-32-9872-9_7

147

based on a series of data including soil organic carbon concentration, bulk density of soil, depth, and duration time. Soil organic carbon stocks and carbon sequestration in urban ecosystems are obviously highly variable.

7.1   Urban Land Use and Soil Change

The global urban population increased from 0.75 billion in 1950 to nearly 4 billion in 2015 (Lal and Stewart 2018). Recently, the United Nations (UN) projects an increase of urban population to 1.35 billion by 2030 and further to nearly 5.35 billion that will be approximately 63% of total world population by 2030 (United Nations 2014). The data suggest that urban areas are expanding and will continue to grow.

At present, the urban area is approximately 340 million ha (Mha) that covers 2.7% of the total land in the world (Seto et al. 2012). The urban area increased by 5.8 Mha from 1970 to 2000 worldwide (Seto et al. 2011) and is projected to increase by an additional 122 Mha in 2030. Changes of the urban land use could lead to significant impact on the ecosystem service because of losses of habitat, biomass, carbon pool in the soils, and biodiversity (Seto et al. 2012). Furthermore, changes of the land vegetation coverage because of changes of land use due to urban expansion also impact the hydrological balance, biogeochemical cycling carbon, and heat balance (Lal and Stewart 2018). Depletion of the carbon pool in the urban area could amount to 0.05 petagrams (Pg) C per year resulting in a total loss of 5% of the total emissions from tropical deforestation (Seto et al. 2012). The soil fertility of farmland is very important to maintain sustainable qualities of agricultural products. Carbon sequestration in soil is one of several alternatives to increase the soil fertility and diversity of farmlands in the urban area. Organic farming utilizes carbon-based amendments, cover crop, green manure, and diverse crop rotations (Reeve et al. 2016) to improve soil physical properties and increase biological availability of soil organic matter among the many other benefits (Reeve et al. 2016). In general, the property of urban soils is not suitable for agricultural plant growth. Most urban soil is highly compacted and contaminated by heavy metals including lead (Pb), copper (Cu), and zinc (Zn), among others. Therefore, the quality of urban soil must be improved for agricultural uses. This chapter addresses a more comprehensive assessment of soil health and carbon sequestration for organic urban farming.

7.2   Soil Health for Organic Agriculture

7.2.1   Concept of Soil Health

"Soil health" is most often defined as "the capacity of a soil to function within ecosystem boundaries to sustain biological productivity, maintain environmental quality, and promote plant and animal health" (Doran and Parkin 1994). In other words, the idea of soil health/quality is related to the capacity of soils to function properly in support of the important ecosystem services needed to sustain soil productivity and maintain environmental quality (Karlen et al. 2001; Nakajima et al. 2015). Thus, a proper soil management is of great importance to the soil sustainability and resiliency for future generations in the city; improper soil managements will lead to adverse changes in soil health. Furthermore, the soil health assessment will enable us to understand future trends in soil physical, chemical, and biological properties as well as its functions overtime, to evaluate whether soil health under different agriculture managements (e.g., organic farming, crop rotation, cover crop, no till-age, etc.) is aggrading, sustaining, or degrading (Karlen et al. 1997; Nakajima et al. 2015). Larson and Pierce (1994) suggested the following three general approaches for assessing soil health:

(1) Comparing management practices for differences in soil health
(2) Comparing the same site over time and establishing trend as a dynamic assessment
(3) Comparing the problem area with non-problem areas within the site

Numerous researchers have proposed several conceptual frameworks or models to assess soil health (Andrews et al. 2004; Shukla et al. 2006; Armenise et al. 2013; Nakajima et al. 2015). However, there is no universally accepted method or tool for evaluating soil health regardless of the many available conceptual frameworks and models proposed.

Some of the methods will be introduced in this chapter. The manual entitled "Comprehensive Assessment of Soil Health (The Cornell Framework) (https://soil-health.cals.cornell.edu/training-manual/) (Moebius-Clune et al. 2016)" has been developed to assist farmers in acquiring appropriate management solutions to build up healthy soils. The assessment specifically measures soil health indicators includ-ing soil physical properties such as soil texture, aggregate stability, available water capacity, and field penetration resistance and soil chemical properties including potentially mineralizable nitrogen, active carbon, and organic matter content, as well as root health, chemical analyses, and macro- and micronutrient level assess-ment. According to the soil test results, prescriptions to improve soil health may include the practice of minimum or no tillage, planting cover crops/green manure, planning crop rotations, preventing damage to wet soil/drainage, and importing organic matter, nutrients, and other amendments (Moebius-Clune et al. 2016). The Framework emphasizes the integration of soil physical, chemical, and biological properties; and the main benefit is the identification of soil physical, chemical, and biological constraints that will prompt farmers to seek improved and more sustain-able soil and crop management practices (Moebius-Clune et al. 2016). In addition, the Cornell Framework provides the identification of feasible management options based on the comprehensive soil test results. If a specific measured soil constraint is identified, appropriate management approaches will be recommended. For exam-ple, if the soil is found to have a low available water holding capacity, the Cornell Framework recommends "adding stable organic materials, such as mulch," "adding compost or biochar," and/or "incorporating high-biomass cover crop" as a short-term management remedy. For a long-term management remedy, the Framework suggests "reducing tillage," "rotating with sod crops," and/or "incorporating high-biomass cover crop" (Moebius-Clune et al. 2016). These suggestions that benefit soil health assessment are based on the database collected nationwide over many years so that a wide range of feasible management options becomes available.

A similar soil health framework (Fig. 7.1) was proposed by Andrews et al. (2004). This soil management assessment framework provides a soil health index ranking from 0 to 1 based on the "scoring function analysis." These values can be easily interpreted to reflect the soil properties under specific situations (Andrews et al. 2004; Nakajima et al. 2015). There are three steps to evaluate the soil health:

(1) Identification of the minimum data set of soil indicators based on management goals, e.g., crop productivity, soil erosivity, etc.: Appropriate soil health indicators are influenced by the soil capacity to perform. Therefore, it is very important to select those indicators.
(2) Indicator interpretation: Each indicator from the minimum data set is transformed into combinable dimensionless scores ranging from 1 to 5 with 5 representing the highest level and 1 representing the lowest level. Generally, there

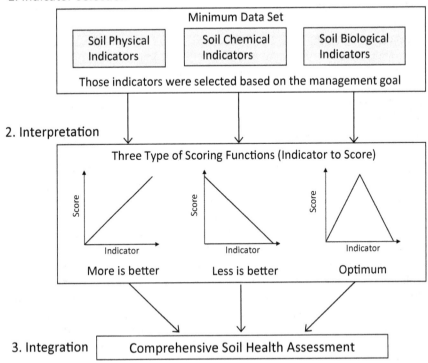

Fig. 7.1 Conceptual framework for scoring function analyses of soil health assessment. (Adapted from Andrews et al. 2004)

are three types of the standard scoring function for soil health: (i) For "More is Better," the level of the indicators is increased in case the soil quality is improving. (ii) For "Less is Better," the line is suitable for decreasing the level of the indicators when the soil health is deteriorating. (iii) For "Optimum," the indicators have an increasingly positive association with soil health up to an optimal level beyond which the soil quality decreases (Fig. 7.1). After selecting an appropriate curve type for indicators, the interpretation needs to be considered.

(3) Integration: The scoring value, e.g., ranging from 1 to 5, of each soil indicator is given specific weights based on their contribution to management goals. Individual scores are multiplied by respective weights and then combined into an overall soil health index.

This method also provides a process for evaluating the ecological function of agriculture soils and identifying individual soil properties that are important to the soil overall condition (Nakajima et al. 2015; Beniston et al. 2016). Currently, conducting a practical assessment of soil health remains a challenging task because it requires an integrated consideration of several soil properties and their spatial and temporal variations.

Nakajima et al. (2016) demonstrated the use of soil quality index for assessing on-farm sites across Ohio and Michigan (Figs. 7.2 and 7.3). This study was conducted at ten on-farm operation sites located in Monroe and Delaware counties in Ohio as well as Dewitt, Clinton, Gladwin, and Osceola counties in Michigan,

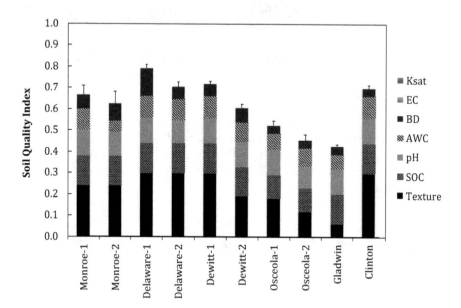

Fig. 7.2 The results of soil quality assessment at ten on-farm sites in Ohio and Michigan. *Ksat* saturated hydraulic conductivity (cm h⁻¹), *EC* electrical conductivity (µS m⁻¹), *BD* bulk density (Mg m⁻³), *AWC* available water capacity (m³ m⁻³), *SOC* soil organic carbon (g kg⁻¹)

Fig. 7.3 Field
measurement for soil
quality assessment in
Columbus, Ohio, USA.
(Photo courtesy of Toru
Nakajima)

USA. The selected sites represent a range of tillage practices (no tillage, minimum tillage, and chisel tillage), and crop rotations (corn-cover crop (C-C), corn-soybean (C-S), and corn-soybean-wheat rotations) are commonly practiced. The selection also considered identifying soil properties that can serve as key indicators in Ohio and Michigan. This research revealed that soil quality index values show a significantly positive correlation with the agronomic yield. In addition, the soil organic carbon and clay content are the key indicators for soil quality index assessment. Overall, the soil quality index is an effective tool for assessing agricultural management.

7.2.2 Soil Physical, Chemical, and Biological Properties

In general, soil health is likely to be linked to the soil organic matter content and soil organic carbon dynamics that directly affect the soil physical, chemical, and biological functions. Soil health assessment requires an evaluation of the soil physical, chemical, and biological attributes to find out how well the natural resources are

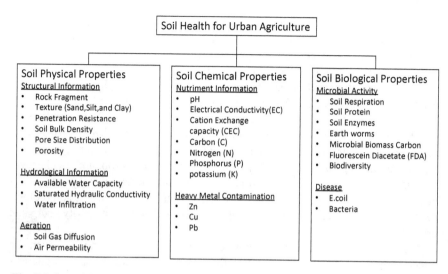

Fig. 7.4 Potential soil indicators including physical, chemical, and biological properties for soil health assessment of organic farming in a city

working (Andrews et al. 2004; Nakajima et al. 2015). Figure 7.4 shows that potential soil indicators for soil health assessment with respect to urban organic farming include the soil physical, chemical, and biological properties.

The soil physical properties (e.g., texture, penetration resistance, bulk density, pore size distribution, porosity, available water capacity, saturated hydraulic conductivity, water infiltration, soil gas diffusion, air permeability, etc.) directly affect the plant root development, water availability, and aeration that in turn influence the plant productivity in agricultural land. Degradation of the soil physical properties such as decreasing water infiltration rate and water holding capacity is more likely to occur in a city. Moreover, the soil moisture condition directly impacts the capability of the plant to uptake nutrients and withstand drought (Pinamonti 1998). In general, organic farming such as applying compost, rotating crop, using cover crops, and practicing the tillage management will improve the soil physical properties. The organic farming practice also decreases the soil bulk density and increases the soil porosity that improves the soil condition to enhance plant root growth and soil aggregate stability as well (Reganold et al. 2001; Nakajima et al. 2015).

Soil chemical properties (e.g., pH, electrical conductivity (EC), cation exchange capacity (CEC), nitrogen (N), phosphorus (P), potassium (K), heavy metal concentrations, etc.) influence nutrient availability, salinity, and alkalinity which directly affect plant/crop productivity and quality. The soil organic matter impacts soil chemical properties and has profound effects on plant growth and yield (Reeve et al. 2016). For example, soil organic matter improves the availability of microelements in the soil as a direct source of slow-release nutrients, elevates the soil buffering capacity and pH, as well as increases soil cation exchange capacity (CEC), which improves plant nutrient availability and decreases leaching potential (Reeve et al. 2016). In addition, measuring the soil toxicity caused by heavy metal deposition (i.e., Pb, Cd, Zn, Ni, and Cu), soil organisms, and vegetation is of great importance for urban organic farming (Beniston et al. 2016).

Soil biological properties and processes influence both soil physical and chemical properties and directly affect the plant health and repeated cultivation damage (Reeve et al. 2016). Soil microorganisms play an important role to break down organic matters contained in the soil, to mineralize and immobilize micronutrients for the plant growing in the soil, and to maintain the soil structure through enhancing the soil aggregates (Fig. 7.5). Applying compost, cover crop, and biochar will increase soil microbial activities and mediate diseases caused by soil-borne pathogens in field trials (Drinkwater et al. 1995; Reeve et al. 2016). Thus, the soil biological properties are crucial to soil health, plant health, and crop productivity in organic urban farming.

There are many benefits of carrying out organic urban farming; however, the effects of organic farming on plant growth and health concerns remain an interesting topic for ongoing and future research.

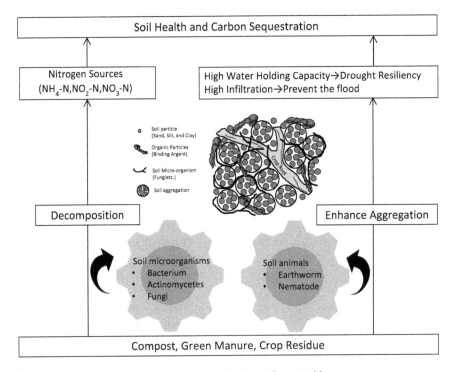

Fig. 7.5 Benefits for applying compost, green manure, and crop residue

7.3 Carbon Sequestration in a City

7.3.1 Carbon Sequestration

The global soil carbon stock amounts to 2500 Pg (P: peta = 10^{15}) including 1500 Pg of soil organic carbon and 950 Pg of soil inorganic carbon (Lal 2004). The soil carbon stock is about 3.3 times the atmospheric carbon stock of 760 Pg and 4.5 times the biotic carbon stock of 560 Pg (Lal 2004). The atmospheric CO_2 concentration has increased by approximately 35% since the start of industrial revolution and is projected to reach 700 ppm by the end of this century (IPCC 2015). Soil is the major stock for terrestrial global carbon. Thus, the concerns of climate change have stimulated interests in utilizing soils for a long-term carbon storage, particularly in an attempt to offset carbon emissions associated with anthropogenic activities (DeLuca and Boisvenue 2012). Furthermore, carbon sequestration is implemented by transferring atmospheric CO_2 into soil organic carbon, and the sequestrated carbon is not easily and immediately released and reemitted into the atmosphere (Lal 2004). Therefore, the carbon sequestration practice is an important operation to mitigate climate change (Lal 2004), whereas soil respiration releases the largest flux of biogeochemical carbon into the atmosphere to contribute to climate change.

The increasing global human population with a proportion of the population residing in urban regions leads to the expansion of urban regions across the world (Lal and Stewart 2018). Urbanization usually causes the accelerated depletion of soil organic stocks. In addition, severe depletion of the soil organic carbon stock degrades soil quality/health, reduces biomass productivity, and adversely impacts water quality (Lal 2004).

7.3.2 Monitoring Carbon Sequestration

Recycling dead and decaying organic matter is essential to storing plant nutrients, denaturing pollutants, filtering water, sequestrating carbon, moderating climate change, storing germplasm, and providing habitat for biodiversity (Lal 2017). The three principal components of soil organic matter are (1) plant and animal residues and living microbial biomass, (2) active or labile soil organic matter, and (3) relatively stable soil organic matter (Lal 2017).

The assessment of carbon stocks in an urban soil is more complicated than assessing the general agriculture soils because the urban soil is exposed to the anthropogenic influence both directly through pollution, sealing, and overcompaction and indirectly through alteration of the soil forming factors (Lal and Stewart 2018). Currently, assessment of the carbon sequestration focuses on natural ecosystems such as forest and meadows and agricultural ecosystems (e.g., Kumar et al. 2012; Post and Kwon 2000; Lal 2005; Nakajima et al. 2016). Still, the carbon sequestration rate or carbon stock in urban regions is less known than for other natural and agricultural ecosystems.

In general, the carbon sequestration rate (C sequestration rate) is determined based on a series of data including soil organic carbon concentration (*SOC*), bulk density of soil (*BD*), depth (*D*), and duration time (*T*):

$$\text{C sequestration rate}\left[M\,L^{-2}\,T^{-1}\right] = BD\left[M\,L^{-3}\right] \times SOC\left[M\,M^{-1}\right] \times D[L] \div T[T] \quad (7.1)$$

Soil bulk density is determined by the ratio of the dry weight of undisturbed soil cores with known dimensions (Grossman and Reinsch 2002) to soil organic carbon concentration that is measured by using the dry combustion method or wet oxidation method (Ellert et al. 2002).

Monitoring carbon sequestration is still an important ongoing research topic, although some researchers have reported some research results on soil organic carbon stocks or carbon sequestration. Table 7.1 shows soil organic carbon stocks and carbon sequestration in urban regions. The above discussions on soil organic carbon stocks and carbon sequestration are based on a limited number of research results reported by researchers. In addition, soil organic carbon stocks and carbon sequestration in urban ecosystems are obviously highly variable. Therefore, more investigations need to be conducted at multiple locations by using precise measurement of soil organic carbon stocks and carbon sequestration.

Table 7.1 Soil organic carbon stocks and carbon sequestration in an urban region

Location	Area (km^2)	Land over/use	SOC stock/carbon sequestration	Reference
New York, USA	2663	Urban forest	15.3 (mg C ha^{-1} year^{-1})	Nowak and Crane (2002)
Illinois, USA	3054	Urban forest	10.0 (mg C ha^{-1} year^{-1})	
Ohio, USA	3800	Urban forest	10.1 (mg C ha^{-1} year^{-1})	
Texas, USA	2790	Urban forest	3.0 (mg C ha^{-1} year^{-1})	
Michigan, USA	2225	Urban forest	9.0 (mg C ha^{-1} year^{-1})	
Arizona, USA	1050	Urban forest	3.0 (mg C ha^{-1} year^{-1})	
Southern California, USA	–	Turf	1.4 (mg C ha^{-1} year^{-1})	Townsend-Small and Czimczik (2010)
Moscow, Russia	–	Turf/forest	28.1 (mg C ha^{-1})	Vasenev and Kuzyakov (2018)
Leicester, UK	–	Turf/forest	86 (mg C ha^{-1})	Edmondson et al. (2014)
Hong Kong, China	–	Turf/forest	12.6 (mg C ha^{-1})	Kong et al. (2014)
Shanghai, China	–	Turf/forest	14.7 (mg C ha^{-1})	
Victoria, Australia		Herbaceous/forest	2.5 (mg C ha^{-1})	Livesley et al. (2010)

References

Andrews SS, Karlen DL, Cambardella CA (2004) The soil management assessment framework. Soil Sci Soc Am J 68(6):1945–1962

Armenise E, Redmile-Gordon MA, Stellacci AM, Ciccarese A, Rubino P (2013) Developing a soil quality index to compare soil fitness for agricultural use under different managements in the Mediterranean environment. Soil Tillage Res 130:91–98

Beniston JW, Lal R, Mercer KL (2016) Assessing and managing soil quality for urban agriculture in a degraded vacant lot soil. Land Degrad Dev 27(4):996–1006

Deluca TH, Boisvenue C (2012) Boreal forest soil carbon: distribution, function and modelling. Forestry 85(2):161–184

Doran JW, Parkin TB (1994) Defining and assessing soil quality. Defining soil quality for a sustainable environment, vol 35. Soil Science Society of America and American Society of Agronomy, Madison, pp 1–21

Drinkwater LE, Letourneau DK, van Bruggen WFA, HC SC (1995) Fundamental differences between conventional and organic tomato agroecosystems in California. Ecol Appl 5(4):1098–1112

Edmondson JL, Davies ZG, McCormack SA, Gaston KJ, Leake JR (2014) Land-cover effects on soil organic carbon stocks in a European city. Sci Total Environ 472:444–453

Ellert BH, Janzen HH, Entz T (2002) Assessment of a method to measure temporal change in soil carbon storage. Soil Sci Soc Am J 66(5):1687–1695

Grossman RB, Reinsch TG (2002) 2.1 Bulk density and linear extensibility. Methods of soil analysis: Part 4 physical methods, (methodsofsoilan 4), pp 201–228

Karlen DL, Mausbach MJ, Doran JW, Cline RG, Harris RF, Schuman GE (1997) Soil quality: a concept, definition, and framework for evaluation (a guest editorial). Soil Sci Soc Am J 61(1):4–10

Karlen DL, Andrews SS, Doran JW (2001) Soil quality: current concepts and applications. Adv Agron 74:1–40

Kong L, Shi Z, Chu LM (2014) Carbon emission and sequestration of urban turf grass systems in Hong Kong. Sci Total Environ 473:132–138

Kumar S, Kadono A, Lal R, Dick W (2012) Long-term no-till impacts on organic carbon and properties of two contrasting soils and corn yields in Ohio. Soil Sci Soc Am J 76(5):1798–1809

Lal R (2004) Soil carbon sequestration impacts on global climate change and food security. Science 304(5677):1623–1627

Lal R (2005) Forest soils and carbon sequestration. For Ecol Manag 220(1–3):242–258

Lal R (2017) Managing urban soils for food security and adaptation to climate change. In: International congress on soils of urban, industrial, traffic, mining and military areas. Springer, Cham, pp 302–319

Lal R, Stewart B (eds) (2018) Urban soils. CRC Press, Boca Raton

Larson WE, Pierce FJ (1994) The dynamics of soil quality as a measure of sustainable management. Defining soil quality for a sustainable environment. Soil Science Society of America and American Society of Agronomy, Madison, pp 37–51

Livesley SJ, Dougherty BJ, Smith AJ, Navaud D, Wylie LJ, Arndt SK (2010) Soil-atmosphere exchange of carbon dioxide, methane and nitrous oxide in urban garden systems: impact of irrigation, fertiliser and mulch. Urban Ecosyst 13(3):273–293

Moebius-Clune BN, Moebius-Clune DJ, Gugino BK, Idowu OJ, Schindelbeck RR, Ristow AJ, van Es HM, Thies JE, Shayler HA, McBride MB, Kurtz KSM, Wolfe DW, Abawi GS (2016) Comprehensive assessment of soil health – the Cornell framework, edition 3.2. Cornell University, Geneva

Nakajima T, Lal R, Jiang S (2015) Soil quality index of a Crosby silt loam in Central Ohio. Soil Tillage Res 146:323–328

Nakajima T, Shrestha RK, Jacinthe PA, Lal R, Bilen S, Dick W (2016) Soil organic carbon pools in ploughed and no-till Alfisols of Central Ohio. Soil Use Manag 32(4):515–524

Nowak DJ, Crane DE (2002) Carbon storage and sequestration by urban trees in the USA. Environ Pollut 116(3):381–389

Pinamonti F (1998) Compost mulch effects on soil fertility, nutritional status and performance of grapevine. Nutr Cycl Agroecosyst 51(3):239–248

Post WM, Kwon KC (2000) Soil carbon sequestration and land-use change: processes and potential. Glob Chang Biol 6(3):317–327

Reeve JR, Hoagland LA, Villalba JJ, Carr PM, Atucha A, Cambardella C, Delate K (2016) Organic farming, soil health, and food quality: considering possible links. In: Advances in agronomy, vol 137. Academic, Amsterdam, pp 319–367

Reganold JP, Glover JD, Andrews PK, Hinman HR (2001) Sustainability of three apple production systems. Nature 410:926–930

Seto KC, Fragkias M, Güneralp B, Reilly MK (2011) A meta-analysis of global urban land expansion. PLoS One 6(8):e23777

Seto KC, Güneralp B, Hutyra LR (2012) Global forecasts of urban expansion to 2030 and direct impacts on biodiversity and carbon pools. Proc Natl Acad Sci 109(40):16083–16088

Shukla MK, Lal R, Ebinger M (2006) Determining soil quality indicators by factor analysis. Soil Tillage Res 87(2):194–204

Townsend-Small A, Czimczik CI (2010) Carbon sequestration and greenhouse gas emissions in urban turf. Geophys Res Lett 37(2). https://doi.org/10.1029/2009GL041675

U.N (2014) World urbanization prospects: the 2014 revision, highlights. Department of Economic and Social Affairs, Population Division, New York

Vasenev V, Kuzyakov Y (2018) Urban soils as hot spots of anthropogenic carbon accumulation: review of stocks, mechanisms and driving factors. Land Degrad Dev 29:1607–1622

Chapter 8
Cover Crop Farming System

Masakazu Komatsuzaki, Takahiro Ito, Tiejun Zhao, and Hajime Araki

Abstract The benefit of applying cover crop depends on farming practices including tillage and cropping systems, as well as land use such as wet paddy or upland fields. The following considerations are some important aspects for evaluating the economic and ecological benefits of cover crops: (1) reducing fertilizer consumption, (2) reducing the use of herbicides, (3) improving crop yields through enhanced soil health, (4) preventing soil erosion caused by wind and water, (5) protecting water quality by mitigating erosion and surface runoff, and (6) attracting beneficial insects and serving as a trap to eliminate harmful insects.

The benefits of using cover crops are generally shown by changes in the soil carbon and nitrogen (N) dynamics resulting from both cover crop ecology and soil biological activities associated with the cover crops. Selection and management of the cover crop strongly influences the amount of carbon input to the soil and carbon release from the crop residue and hence the ability to replenish the soil organic carbon (SOC) pool. Adding cover crop residue in soil can enhance SOC that is closely associated with soil microbial diversity and activity. Changes of microbial quantity and quality in the soil indicate changes in soil quality caused by the use of cover crops. These changes are important responsive indicators of how the soil management practices affect crops. The no-tillage system is another alternative that has been used increasingly for crop production because of its significant environmental advantages over moldboard plow. Hairy vetch (HV), a cover crop in greenhouse, has a superior capability in supplying nitrogen because of its excellent efficiency in fixing nitrogen biologically. Because intensive tomato production in greenhouse needs a significant quantity of nutrition for a healthy growth and sufficient yield, utilizing cover crops is desirable to reduce the consumption of chemical fertilizer.

M. Komatsuzaki (✉)
College of Agriculture, Ibaraki University, Ami, Japan
e-mail: masakazu.komatsuzaki.fsc@vc.ibaraki.ac.jp

T. Ito · T. Zhao
Niigata Agro-Food University, Niigata, Japan
e-mail: takahiro-ito@nafu.ac.jp

H. Araki
Field Science Center for Northern Biosphere, Hokkaido University, Hokkaido, Japan

© Springer Nature Singapore Pte Ltd. 2020
S. Tojo (ed.), *Recycle Based Organic Agriculture in a City*,
https://doi.org/10.1007/978-981-32-9872-9_8

Developing and implementing ecospecific, eco-friendly, and system-based soil management methods are necessary to deal with the growing demands and pressures on land and water resources.

8.1 Significance of Cover Crop in Agroecosystem

Cover crop will be a key element to develop the recycle-based organic farming in a city known as the conservation agriculture. More than 2000 years ago, farmers in China and Mediterranean nations sowed cover crops into soil to improve the soil productivity. Chinese milk vetch (*Astragalus sinicus* L.) was the most useful cover crop in China for more than 2000 years: it was introduced to Japan from China around the seventh century. Applying green manure and cover crops was noted as one of the most important tools for improving soil fertility in "Nougyou Zensyo," which is a Japanese corpus of agricultural techniques of the Edo era published in 1697 (Yasue 1993).

The benefit of applying cover crop also depends on farming practices including tillage and cropping systems, as well as land use such as wet paddy or upland (dry) fields. The following considerations are some important aspects for evaluating the economic and ecological benefits of cover crops: (1) reducing fertilizer consumption by contributing nitrogen (N) from cover crop and by scavenging and mining soil nutrients (Horimoto et al. 2002; Komatsuzaki 2002); (2) reducing the use of herbicides because many cover crops behave as smother plants to suppress weeds effectively by outcompeting weeds for water and nutrients – some cover crop residues or growth of leaf canopy blocks light, alters the frequency of light waves, and changes soil surface temperature in addition to being a source of root exudates or compounds to provide natural herbicidal effects (Creamer et al. 1996; Fujii et al. 2004); (3) improving crop yields through enhanced soil health resulting from the addition of organic matter that stimulates beneficial soil microbes and enhances the nutrient cycle (Gu et al. 2004); (4) preventing soil erosion caused by wind and water because the aboveground portion of cover crops quickly covers the soil surface (Dabney 1998); (5) protecting water quality by mitigating erosion and surface runoff because cover crops scavenge soil residual nutrients, thereby reducing nutrients and agricultural chemicals contained in non-point agricultural discharges (Staver and Brinsfield 1998; Delgado 1998); and (6) attracting beneficial insects and serving as a trap to eliminate harmful insects and/or providing food and shelter for wildlife (Schutter and Dick 2002). These significant environmental and productivity benefits vary depending on not only location and season (Sarrantonio 1998a, b) but also land use practices.

One of the attractive aspects of cover crops, especially in urban farming, is to provide an adequate soil cover to prevent wind erosion. In the Kanto region, Japan, a strong monsoon called *kara-kaze* in the Japanese language often occurs from winter to early spring. This strong and dry wind removes soil moisture and exposes the soil to erosion when the field is left fallow after the summer crop harvest. Wind erosion physically removes the most fertile part of the soil such as organic matter, clay, and silt, thus causing declined soil productivity. Some soil particles from the eroded field become suspended in the atmosphere. This airborne dust obscures visibility, pollutes the air, fills road ditches, adversely impacts water quality, causes automobile accidents, fouls machinery, and imperils human health. Thus, wind erosion threatens not only the sustainability of the land but also the viability and quality of human life for urban communities (Komatsuzaki and Suzuki 2008). Figure 8.1 shows the comparison of the agroecosystems practicing cover crop and the fallow field after crop harvest in winter until spring. For the fallow field, soil wind erosion is quite significant in winter, and N leaching becomes serious in early spring when the precipitation is abundant while evaporation of the soil moisture is limited due to low atmospheric temperature. For the cover crop field, however, the cover crop vegetation provides a soil cover to prevent wind erosion and keep soil moisture, enhances soil organic matter, and scavenges the residual N in the soil. Cover crop management is carried out through selection of crop spices, as well as adjustment of the seeding time, seeding rate, and seeding termination time because these factors strongly affect the plant biomass and significantly alleviate wind erosion (Komatsuzaki and Suzuki 2008).

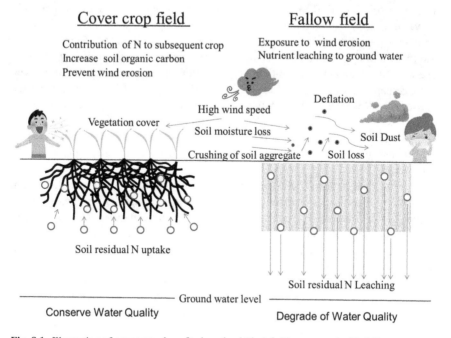

Fig. 8.1 Illustration of cover crop benefits in upland (dry) fields compared with fallow treatment. The fallow period field suffers serious wind soil erosion and nutrient leaching to groundwater, whereas the cover crop field conserves the surface soil due to the use of soil cover and enhances nutrient uptake by the crop that also contributes to nutrient scavenge

8.2   Carbon and Nitrogen Cycle in Upland Field

The benefits of using cover crops are generally shown by changes in the soil carbon and N dynamics resulting from both cover crop ecology and soil biological activities associated with the cover crops. These effects are manifested primarily in the chemical composition and physical and ecological characteristics of the cover crops. Selection and management of the cover crop strongly influences the amount of carbon input to the soil and carbon release from the crop residue and hence the ability to replenish the soil organic carbon (SOC) pool. The tillage system is also an important factor for promoting the decomposition of cover crop in the soil. Research is underway to evaluate the use of fall-planted cover crops such as rye and hairy vetch in various tillage systems in the Kanto region of Japan (Higashi et al. 2014).

Table 8.1 lists data on accumulations of dry matter (DM) and nutrient for various cover crops available in Japan. Based on ecological and economic principles, these systems assist in maintaining agriculture sustainability because they are more efficient in using solar energy and other natural resources, maintaining ecological equilibrium, controlling epidemics of pests and weeds, and improving agricultural productivity.

Cover crop residues replenish the organic matter in soil where it is naturally decomposed. Returning more cover crop residues to the soil provides more organic matter in soil. Magdoff and Van Es (2000) carried out a 5-year experiment with clover in California and demonstrated that cover crops increase the organic matter content in the top 5 cm layer of soil from 1.3 to 2.6% and in the 5–15 cm layer from 1.0 to 1.2%. Higashi et al. (2014) evaluated the capabilities of cover crops and tillage systems to enhance the carbon sequestration based on the results of a 9-year field research conducted at Ibaraki University's Center for International Field Agriculture Research and Education. The soil is a Humic Allophane soil (Haplic Andsolos) that is common to the Kanto region of Japan. The plot with cover crop shows significant increase of soil organic carbon in the top 0–30 cm layer of the no-till and rotary-till soil, when comparing with winter fallow plots without cover crop.

Cover crops can also reduce non-point source pollution caused by nutrients and agricultural chemicals contained in surface runoff from agricultural fields. In particular, non-legume cover crops have been shown to prevent excess N from accumulating in the soil of upland dry field and leaching to groundwater (Komatsuzaki 2009; Komatsuzaki and Wagger 2015; Komatsuzaki 2017). In the Kanto region, contamination of groundwater by N from fertilizer is of concern in fall and spring when intensive precipitation and relatively low evaporation cause relatively high N leaching. For example, during the fallow season, nutrients, such as nitrate (NO_3^-), sometimes leach from a fertile soil into groundwater (Miura and Ae 2006). Nishio (2001) reported that the nitrate concentration in groundwater often exceeds 10 ppm and the highest concentration observed in a vegetable production area goes above 77 ppm. Production of most vegetable and field crops occurs during the period between spring and summer as well as between autumn and winter. This results in

Table 8.1 Dry matter and nutrient accumulation of typical cover crop based on various studies carried out in Japan

Cover crop	Advantage	Dry matter Mg ha⁻¹	C/N ratio	N kg ha⁻¹	P_2O_5 kg ha⁻¹	K_2O kg ha⁻¹	Source
Winter cover crop							
Chinese milk vetch (*Astragalus sinicus* L.)	N contribution	3–6	15	70–150	10–30	50–100	Kanagawa Prefecture Office (2004)
Hairy vetch (*Vicia villosa* Roth)	Weed suppression	6	18	160	50	70	Kanagawa Prefecture Office (2004)
		3–5	–	–	–	–	Higashi et al. (2014)
Subterranean clover (*Trifolium subterraneum*)	N contribution	5–6	38	51–65	–	–	Komatsuzaki (2002)
Rye (*Secale cereale* L.)	Improve soil organic matter N scavenging	3–7	35	44–87	–	–	Komatsuzaki (2009)
		3–12	–	–	–	–	Higashi et al. (2014)
Italian ryegrass (*Lolium multiflorum* Lam.)	Improve soil organic matter	4–6	20	100–200	10–40	200–400	Kanagawa Prefecture Office (2004)
		2–4	17	90	–	–	Komatsuzaki (2004)
Oat (*Avena sativa* L.)	Improve soil organic matter Inhibit nematode	8	20	200	30	350	Kanagawa Prefecture Office (2004)
		1.1–1.5	13	27–45	–	–	Komatsuzaki (2004)
Wheat (*Triticum aestivum* L.)	Improve soil organic matter N scavenges	1.5–2.0	19	30–42	–	–	Komatsuzaki (2004)
Summer cover crop							
Sorghum (*Sorghum bicolor*)	Improve soil organic matter N scavenges	10–30	35	200–300	30–50	300–700	Kanagawa Prefecture Office (2004)
		2–20	20–36	83–483	5–40	40–341	Komatsuzaki et al. (2012)

(continued)

Table 8.1 (continued)

Cover crop	Advantage	Dry matter Mg ha^{-1}	C/N ratio	N kg ha^{-1}	P$_2$O$_5$ kg ha^{-1}	K$_2$O kg ha^{-1}	Source
Guinea grass (*Panicum maximum*)	Improve soil organic matter Inhibit nematode N scavenging	10	18	200	70	350	Kanagawa Prefecture Office (2004)
Sunn hemp (*Crotalaria juncea* L.)	Inhibit nematode	5	40	100	30	170	Kanagawa Prefecture Office (2004)
Marigold (*Tagetes patula*)	Inhibit nematode	7	17	190	–	–	Kanagawa Prefecture Office (2004)
Crotalaria (*Crotalaria spectabilis* Roth)	N contribution	1–4	13–15	18–120	1–8	9–52	Komatsuzaki et al. (2012)
Sesbania (*Sesbania cannabina* Pers.)	N contribution Subsoil crushed	0.1–2	15–37	4–25	0.3–5	2–15	Komatsuzaki et al. (2012)

relatively sparse vegetation cover in the fields during early spring and early autumn when nutrient leaching is observed to be the most prominent.

However, applying winter non-legume cover crops may be an effective method for managing inorganic N in soils because these crops are capable of restoring soil residual N after summer crop is harvested (Lacroix et al. 2005). Several non-legume winter cereal species, including rye, wheat, and oat, are recommended as cover crops in Japan (Komatsuzaki 2004); and these non-legume cover crops are known to scavenge residual soil N (Wagger and Mengel 1988; Wagger 1989). Figure 8.2 shows the yearly soil inorganic N concentration at the 60–90 cm soil depth layer. This suggests that non-legume cover crop is effective to prevent the soil N from leaching into deeper soil layers. The same non-legume cover crops are also known to be used in rotation with summer crops for many years in the Kanto region. Such a system works well for corn (*Zea mays* L.), tobacco (*Nicotiana tabacum* L.), peanuts (*Arachis hypogaea*), and soybean (*Glycine max.* Merr.) because these non-legume cover crops have wider ranges of planting time than other legume cover crops (Wagger 1989). The potential for recovering N is strongly influenced by factors relating to cover crop species and crop management (Ditsch and Alley 1991). The most critical period to recover residual N following a summer crop begins immediately after the harvest until the cover crop reaches winter growth dormancy. Therefore, early sowing of cover crop is essential for maximizing N recovery.

Komatsuzaki and Wagger (2015) reported that accumulation of N in cover crop increases with a delay at the growth termination date, although the interaction

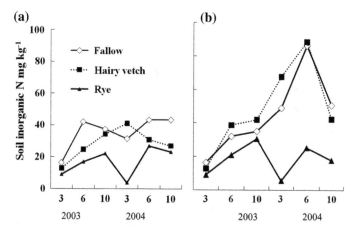

Fig. 8.2 Changes in soil inorganic N concentration at 60–90 cm depth layer in relation to cover cropping. Cover crops were planted following dry field rice cultivation in no- tillage (**a**) and conventional-tillage (**b**) systems. Experiments were conducted at the Ibaraki University Experimental Farm in Japan. (Data from Komatsuzaki and Mu 2005)

between cover crop species and the planting date varies. October planting of rye and triticale shows greater N accumulation with a March growth termination date, averaging 37.8 kg N ha^{-1} for rye and 37.6 kg N ha^{-1} for triticale. These values are 7.8% and 14.1% higher than those of black oat and wheat, respectively, at the same planting and growth termination dates. Wheat planted in November shows the highest N accumulation (average 57.9 kg N ha^{-1}) at the late April/early May termination date; it is 3.8–7.9% higher than that of other plant species. In contrast, black oat (*Avena strigosa* Schreb) shows a greater capability to scavenge 58.7 kg N ha^{-1} soil residual N on average with a combined late planting and growth termination dates.

Cover crops have the benefits with respect to N dynamics in which the cover crop recovers the residual N; later, the N contributes to the growth of subsequent crop. The issue on how much N can be retained and later supplied to subsequent crops is especially difficult to estimate because of N immobilization in soil easily. The farming method after those legume cover crops are terminated affects the amount and timing of N release from the legume plant to the soil. Other environmental factors also contribute significantly to the rate of N released from the cover crop, and much of the N released is directly controlled artificially in the field (Sarrantonio 1998a, b; Magdoff and Van Es 2000). The soil management practice also influences the quantity of nutrients removed from the soil by the cover crops and the amount of nutrients available to the subsequent crop after the cover crops decompose. Plowing cover crops under in early spring will increase the decomposition rate of the cover plant material. Allowing the cover crop to mature will improve the nutrient accumulation in the cover crop biomass, but this may reduce the microbial capability to decompose organic residues for a short-term application. Leaving the cover crop on the soil surface also reduces the rate of decomposing the organic matter as compared

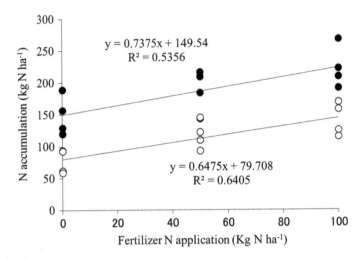

Fig. 8.3 Subterranean clover N contribution to subsequent silage corn compared with fallow treatments. ● indicates subterranean clover-corn rotation, and ○ indicates fallow-corn rotation. (Data from Komatsuzaki 2002)

with the plowing practice. Thus, a mature cover crop will benefit subsequent crops by allowing more residual organic matter to accumulate on the soil surface.

Sarrantonio (1998a, b) reported that the amount of N supplied by cover crops to subsequent crops depends on the tillage system. Komatsuzaki (2004) estimated that the nitrogen released from the residue of cover crops and uptaken by corn can be estimated by examining the difference in N uptake for the corn harvested between a sub-clover reseeding system and a fallow system. The amount of N added by the subterranean clover to the subsequent corn as fertilizer was estimated to be 48–75 kg N ha^{-1} (Fig. 8.3). This value was somewhat higher than the N accumulated in clover during spring, because the subterranean clover N accumulation ranges from 51.1 to 65.1 kg N ha^{-1}. However, these values reflect only the aboveground portion of the plant; they do not include N accumulated in plant roots.

8.3 Soil Biodiversity and Ecosystem Services

Healthy soil is defined as a stable soil system with strong resilience to stress, high biological diversity, and high levels of internal nutrient cycling (van Bruggen and Semenov 2000). Major microbial activities in soil include decomposition of organic matter, mineralization of nutrients, fixation of nitrogen, and suppression of crop pests and protection of roots. However, some microbes in soil may be parasitic to cause injury to plants. Thus, the practice to improve soil quality needs to address this issue in order to maintain the soil health. Practicing appropriate management of cover crops in croplands may significantly alleviate the parasitic microbial issue, but this practice depends on understanding the basic microbiology.

Adding cover crop residue in soil can enhance SOC that is closely associated with soil microbial diversity and activity. Changes of microbial quantity and quality in the soil may indicate changes in soil quality caused by the use of cover crops. These changes are important responsive indicators of how the soil management practices affect crops. Soil microbes may react differently to adapt to the various crop residues in different tillage systems. For example, applying leguminous cover crops significantly increases the soil microbial populations in no-tillage field. However, the increase of microbial population is not as obvious using rye cover crops and also in all rotary- and plow-tillage systems. On the other hand, significantly higher fungal biomass exists in no-till field with cover crops in May and October. This suggests that growing cover crops with no-till methods leads to a different soil microbial community and structure as compared with conventional tilled soil ecosystems (Zhaorigetu et al. 2008).

Cover crops in tillage systems also show significant effects on nematode communities. Ito et al. (2015a) reported that no-tillage plots with hairy vetch cover crops have the highest total nematode population among all plots. However, the population of *Heterodera* spp., which are pathogen nematodes for field rice production, is significantly lower in no-tillage fields than in plow- and rotary-tillage fields.

Cover crops and tillage systems also showed significant effects on soil fauna populations and species. The no-tillage system with cover crop treatments has significantly more soil macrofauna populations and species, whereas plow- and rotary-tillage systems with cover crop do not show increase of soil fauna populations. The soil ecosystem for no-tillage plots with rye cover crop shows a significantly higher correlation between the predator population and soil mesofauna than other systems although this correlation was poor in soil ecosystems for conventional tillage systems (Ito et al. 2015b). These results suggest that soil ecosystems in no-tillage fields with cover crops improve the soil food web structure (Fig. 8.4).

The addition of cover crop residue to soil can also increase the populations of soil microbes, micro- and macro-arthropods, and earthworms, all of which contribute to efficient nutrient cycling and help improve the soil structure (Arai et al. 2013). The interaction between tillage and cover crop on microbial biomass and their activities is also significant. Zhaorigetu et al. (2008) reported that the combined practice of no tillage with rye cover crop treatment increases the fungal biomass but not bacterial populations in the top 0–10 cm layer of soils. Such increase in fungal biomass is not found in 10–20 cm and 20–30 cm depth of the cover-cropped no-tillage soil.

Increasing fungus activity can enhance the soil aggregate stability. Changing soil microbial communities and adding organic matter transfer microbial communities into newly evolved biological consortia with distinctive functions to provide selective environmental advantages (Nakamoto et al. 2012). Fukui (2003) reported that the incorporation of cover crops assists soils in suppressing various soil-borne pathogens such as *Pythium*, *Phytophthora*, and *Rhizoctonia*. Soil enrichment with

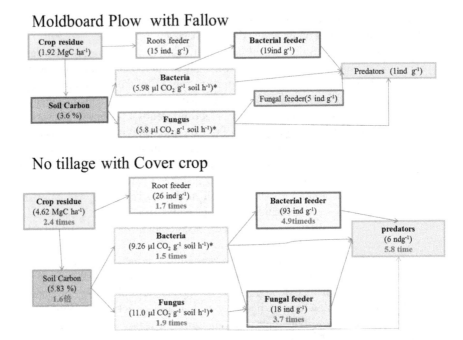

Fig. 8.4 Comparison of soil ecosystem between moldboard plow with fallow field and no tillage with rye cover crop field. Cover crop C input and soil carbon data were obtained from Higashi et al. (2014); soil bacteria and fungus biomass were obtained from Nakamoto et al. (2012); soil nematode data was obtained from Ito et al. (2015a)

organic matter may be the most fundamental and sustainable approach to the biological control of soil-borne diseases, especially in organic farming.

8.4   No Tillage with Cover Crop Management

The no-tillage system is another alternative that has been used increasingly for crop production in the USA, Europe, South America, and Asia during the last decade because of its significant environmental advantages over moldboard plow. For example, the plot in no-tillage systems has become popular steadily all over the world, and it is now being adopted on more than 95 million ha worldwide. East Asian countries including Japan and China just started to adopt no-tillage farming; however, a number of researches carried out on no-tillage practices show great potential to improve soil quality (Nakamoto et al. 2012; Higashi et al. 2014). The combination of cover crops and no-tillage practice augments the environmental benefits such as reducing N leaching, increasing soil organic matter, and improving soil biological diversity (Komatsuzaki and Ohta 2007).

Fig. 8.5 No-tillage seeder performance under cover crop residue mulch

The technique of seeding cash crop with no tillage by leaving the cover crop residue on the soil surface is still being studied in some Asian countries (Fig. 8.5). Zhao et al. (2012) revealed that the no-tillage seeder with rye consumes more energy than hairy vetch and mixed seeding; the energy consumption increases with higher quantities of cover crop residue. Furthermore, the termination and management of cover crop significantly affect the cover crop biomass accumulation and power consumption of the no-tillage seeder. The results reveal that late cover crop termination increases the biomass while the performance of the no-tillage seeder declines. The conclusion is that flail mower cutting and mixed seeding are appropriate for improving the no-tillage seeder in soybean production.

8.5 Cover Crop for Greenhouse Vegetable Production

More than 70% of fruit vegetables including tomato grown in greenhouse are repeatedly produced with the application of significant amount of chemical fertilizer in Japan. Salt accumulation and injury by continuous cropping are serious problems. Excessive N input causes extra N accumulation in the soil, thus causing serious environmental pollution issues such as the occurrence of nitrate leaching and greenhouse gas emission. For achieving sustainable greenhouse production, applying adequate amount of fertilizer is of great importance.

Hairy vetch (HV), a cover crop in greenhouse, has a superior capability in supplying N because of its excellent efficiency in fixing N biologically (Seo et al. 2006; Almeida Acosta et al. 2011). HV has also other advantages such as easily adapting to low temperature, resisting pests, delaying senescence, covering ground surface effectively, and fitting most vegetable productions especially when it is in rotation with tomatoes (Kumar et al. 2005). Because intensive tomato production in greenhouse needs a significant quantity of nutrition for a healthy growth and sufficient yield, utilizing cover crops is desirable to reduce the consumption of chemical fertilizer. Hajime et al. (2009) estimated that using HV as a cover crop will reduce 50%

the recommended chemical N fertilizer application without any compromise of the fruit yield in tomato production. Practicing HV cultivation at the seeding density of 20–50 kg ha^{-1} before tomato planting and applying the slow-release N fertilizer cause the reduction of N fertilizer application while achieving a normal yield of tomato grown in plastic greenhouse (Sugihara et al. 2014).

Some studies show that the recovery rate of N derived from HV applied in the previous year was from 2.3% to 3.8% in maize and oats, respectively (Almeida Acosta et al. 2011; Cueto-Wong et al. 2001; Seo et al. 2006). Evaluating the recovery rate of HV-derived N under greenhouse conditions is important to indicate decomposition of the incorporated HV under suitable soil temperature and soil moisture. Results of the study conducted in Hokkaido, Japan, show that 50% of HV-derived N is absorbed by tomato plants primarily in 4 weeks after transplanting; the absorbed nitrogen is then distributed into low-position fruits, leaves, and stems. HV-derived N is probably available as alternate fast-release fertilizer in the current year. However, the recovery rate of N derived from HV applied in the previous year was about 4% of total N absorbed by tomato plant during the next year (Sugihara et al. 2016). The results suggest that HV may be used for not only a short-term N source during the current year but also a long-term N source the same as other organic materials for the subsequent year.

8.6   Challenges in Organic Agriculture

In conclusion, like no other human enterprises, the agricultural industry dominates land and water usage with the agroecosystem providing critical products to sustain human life and activities. The concern for soil management will become more widespread and noticeable because increasing agricultural outputs are essential for promoting equity and maintaining global political and social stability. Farmers, consumers, researchers, and policymakers of the world begin to develop and promote recycle-based organic agriculture. Results of intensive research completed in recent years reveal that the practice of adopting cover crop in farming can help mitigate global warming, conserve biodiversity, and maintain soil fertility and productivity. However, these farming practices often do not provide enough return to farmers directly. Therefore, political and social incentives will be required based on the common understanding that healthy soils and agroecosystems are essential for developing a sustainable agriculture. To deal with the growing demands and pressures on land and water resources, developing and implementing ecospecific, ecofriendly, and system-based soil management methods are necessary. Organic agriculture in a city is a challenging task; covering the soil surface with cover crop vegetation will provide multiple benefits to the agriculture practiced in the urban region.

References

Almeida Acosta JA, Amado TJC, Neergaard A, Vinther M, Silva LS, Silveira Nicoloso R (2011) Effect of 15n-labeled hairy vetch and nitrogen fertilization on maize nutrition and yield under no-tillage. Rev Bras Ciênc Solo 35:1337–1345

Arai M, Tayasu I, Komatsuzaki M, Uchida M, Shibata Y, Kaneko N (2013) Changes in soil aggregate carbon dynamics under no-tillage with respect to earthworm biomass revealed by radiocarbon analysis. Soil Tillage Res 126:42–49

Creamer NG, Bennett MA, Stinner BR, Cardina J, Regnier EE (1996) Mechanisms of weed suppression in cover crop-based production systems. HortScience 31:410–413

Cueto-Wong JA, Guldan SJ, Lindemann WC, Remmenga MD (2001) Nitrogen recovery from 15N-labeled green manures: I. recovery by forage sorghum and soil one season after green manure incorporation. J Sustain Agric 17:27–42

Dabney SM (1998) Cover crop impacts on watershed hydrology. J Soil Water Conserv 53:207–213

Delgado JA (1998) Sequential NLEAP simulations to examine effect of early and late planted winter cover crops on nitrogen dynamics. J Soil Water Conserv 53:241–244

Ditsch D, Alley M (1991) Nonleguminous cover crop management for residual N recovery and subsequent crop yields. Journal of Fertilizer Issues 8(1):6–13

Fujii Y, Shibuya T, Nakatani K, Itani T, Hiradate S, Parvez MM (2004) Assessment method for allelopathic effect from leaf litter leachates. Weed Biol Manag 4:19–23

Fukui R (2003) Suppression of soilborne plant pathogens through community evolution of soil microorganisms. Microbes Environ 18:1–9

Gu S, Komatsuzaki M, Moriizumi S, Mu Y (2004) Soil nitrogen dynamics in relation to cover cropping. Jpn J Farm Work Res 39:9–16

Hajime A, Hane S, Hoshino Y, Hirata T (2009) Cover crop use in tomato production in plastic high tunnel. Hortic Environ Biotechnol 50:324–328

Higashi T, Yunghui M, Komatsuzaki M, Miura S, Hirata T, Araki H, Kaneko N, Ohta H (2014) Tillage and cover crop species affect soil organic carbon in andosol, Kanto, Japan. Soil Tillage Res 138:64–72

Horimoto S, Araki H, Ishimoto M, Ito M, Fujii Y (2002) Growth and yield of tomatoes in hairy vetch-incorporated and-mulched field. Jpn J Farm Work Res 37:231–240

Ito T, Araki M, Higashi T, Komatsuzaki M, Kaneko N, Ohta H (2015a) Responses of soil nematode community structure to soil carbon changes due to different tillage and cover crop management practices over a nine-year period in Kanto, Japan. Appl Soil Ecol 89:50–58

Ito T, Araki M, Komatsuzaki M (2015b) No-tillage cultivation reduces rice cyst nematode (*Heterodera elachista*) in continuous upland rice (*Oryza sativa*) culture and after conversion to soybean (*Glycine max*) in Kanto, Japan. Field Crop Res 179:44–51

Komatsuzaki M (2002) New cropping strategy to reduce chemical fertilizer application to silage corn production using subterranean clover reseeding. Jpn J Farm Work Res 37:1–11

Komatsuzaki M (2004) Use of cover crops in upland fields. Jpn J Farm Work Res 39:157–163

Komatsuzaki M (2009) Nitrogen uptake by cover crops and inorganic nitrogen dynamics in andisol paddy rice field. Jpn J Farm Work Res 44:201–210

Komatsuzaki M (2017) Cover crops reduce nitrogen leaching and improve food quality in an organic potato and broccoli farming rotation. J Soil Water Conserv 72:539–549

Komatsuzaki M, Mu Y (2005) Effects of tillage system and cover cropping on carbon and nitrogen dynamics. Ibaraki, Japan, pp 62–67. In: Proceedings and abstracts of ecological analysis and control of greenhouse gas emission from agriculture in Asia. September 2005

Komatsuzaki M, Ohta H (2007) Soil management practices for sustainable agro-ecosystems. Sustain Sci 2:103–120

Komatsuzaki M, Suzuki K (2008) Mitigation of wind erosion adopting grass cover crops in Chinese cabbage production. Jpn J Farm Work Res 43:187–197

Komatsuzaki M, Wagger MG (2015) Nitrogen recovery by cover crops in relation to time of planting and growth termination. J Soil Water Conserv 70:385–398

Komatsuzaki M, Moriizumi S, Gu S, Abe S, Mu Y (2004) A case study on paddy culture utilizing cover cropping. Jpn J Farm Work Res 39:23–26

Komatsuzaki M, Suganuma K, Araki H (2012) Evaluation of growth of summer cover crops and their function for agro-ecosystems based on multivariate analysis. Jpn J Farm Work Res 47:55–65

Kumar V, Abdul-Baki AA, Anderson JD, Mattoo AK (2005) Cover crop residues enhance growth, improve yield, and delay leaf senescence in greenhouse-grown tomatoes. HortScience 40:1307–1311

Lacroix A, Beaudoin N, Makowski D (2005) Agricultural water nonpoint pollution control under uncertainty and climate variability. Ecol Econ 53:115–127

Magdoff F, Van Es H (2000) Building soils for better crops. Sustainable Agriculture Network, Beltsville

Miura N, Ae N (2006) Possibility of leaching of organic nitrogen in a field under heavy application of organic matter-model experiment using soil columns. Soil Sci Plant Nutr 52:134–135

Nakamoto T, Komatsuzaki M, Hirata T, Araki H (2012) Effects of tillage and winter cover cropping on microbial substrate-induced respiration and soil aggregation in two Japanese fields. Soil Sci Plant Nutr 58:70–82

Nishio M (2001) A method to assess the risk of nitrate pollution of groundwater by the nitrogen fertilization load from the individual crop species. Jpn J Soil Sci Plant Nutr 72:522–528

Kanagawa Prefecture Office (2004) Use of green manure. http://www.pref.kanagawa.jp/docs/f6k/cnt/f6802/documents/848439.pdf

Sarrantonio M (1998a) Building fertility and tilth with cover crops. Sustainable Agriculture Network, Maryland

Sarrantonio M (1998b) Building fertility and tilth with cover crops. In: Managing cover crops profitably. Sustainable Agriculture Network, cop., Beltsville, pp 16–24

Schutter ME, Dick RP (2002) Microbial Community profiles and activities among aggregates of winter fallow and cover-cropped soil, Published as Paper No. 11590 of the Oregon Agric. Exp. Stn., Oregon State Univ., Corvallis, OR. Soil Sci Soc Am J 66:142–153

Seo JH, Meisinger JJ, Lee HJ (2006) Recovery of nitrogen-15–labeled hairy vetch and fertilizer applied to corn. Agron J 98:245–254

Staver KW, Brinsfield RB (1998) Using cereal grain winter cover crops to reduce groundwater nitrate contamination in the mid-Atlantic coastal plain. J Soil Water Conserv 53:230–240

Sugihara Y, Ueno H, Hirata T, Araki H (2014) Hairy vetch derived-N uptake by tomato grown in a pot containing fast-and slow-release N fertilizer. J Jpn Soc Hortic Sci 83:222–228

Sugihara Y, Ueno H, Hirata T, Komatsuzaki M, Araki H (2016) Contribution of N derived from a hairy vetch incorporated in the previous year to tomato N uptake under hairy vetch-tomato rotational cropping system. Hortic J 85:217–223

van Bruggen AHC, Semenov AM (2000) In search of biological indicators for soil health and disease suppression. Appl Soil Ecol 15:13–24

Wagger MG (1989) Cover crop management and nitrogen rate in relation to growth and yield of no-till corn. Agron J 81:533–538

Wagger M, Mengel D (1988) The role of non leguminous cover crops in the efficient use of water and nitrogen 1. In: Cropping strategies for efficient use of water and nitrogen, pp 115–127

Yasue T (1993) Chinese milk vetch. Japan. Nou-bun-Kyo, Tokyo

Zhao T, Zhao Y, Higashi T, Komatsuzaki M (2012) Power consumption of no-tillage seeder under different cover crop species and termination for soybean production. Eng Agric Environ Food 5:50–56

Zhaorigetu KM, Sato Y, Ohta H (2008) Relationships between fungal biomass and nitrous oxide emission in upland rice soils under no tillage and cover cropping systems. Microbes Environ 23:201–208

Kanagawa Prefecture Office (2004) Use of green manure. http://www.pref.kanagawa.jp/docs/f6k/cnt/f6802/documents/848439.pdf

Chapter 9
Symbiotic Coexistence of Paddy Field and Urban Ecosystem

Takashi Motobayashi and Seishu Tojo

Abstract The paddy field is primarily used as an agricultural system to produce rice, which is the main staple food in Japan. The paddy field plays an important role in the various ecosystem functions such as flood control, groundwater recharge, improvement of water quality, local climate mitigation, fish culture and other non-rice productions, fostering culture and landscape, and maintaining biodiversity, among others. Paddy fields provide an important habitat to foster biodiversity. The remaining rice paddies in urban regions have an important role to alleviate the deterioration of the urban environment.

In the first section, no-tillage cultivation studies carried out so far to investigate its effects on biodiversity in paddy fields are explained. Results show that spiders and carabid beetles prefer no-tillage soils in which the litter accumulate to the soil surface where abundant decomposers exist. Actually, in various types of crops, the density of soil-inhabiting predators such as spiders and carabid beetles is known to be higher in no-tillage or reduced-tillage cultivation field. The results also suggest that predation by spiders (especially lycosid spiders) is an important mortality factor for older larvae of the straight swift, and the effect is greater for the no-tillage paddy fields than the conventional paddy fields.

In the following sections, an integrated rice farming method that uses crossbred ducks being practiced since 150 years ago, known as Aigamo farming in Japanese, is described. The rice-duck farming is a promising technique for producing organic rice in Southeast Asia. The farming system involves organic food certification systems, organic farmers' cooperatives, community-wide organic farming, localized technical extension and educational services, and integration of farms and rice-duck. An experiment undertaken for the simultaneous cultivation of rice plants and baby crossbred ducks at the university farm in Fuchu, Tokyo, shows the effects of

T. Motobayashi (✉)
Field Science Center, Tokyo University of Agriculture and Technology, Fuchu, Tokyo, Japan
e-mail: takarice@cc.tuat.ac.jp

S. Tojo
Institute of Agriculture, Tokyo University of Agriculture and Technology, Fuchu, Tokyo, Japan
e-mail: tojo@cc.tuat.ac.jp

© Springer Nature Singapore Pte Ltd. 2020
S. Tojo (ed.), *Recycle Based Organic Agriculture in a City*,
https://doi.org/10.1007/978-981-32-9872-9_9

the release of crossbred ducks on the growth and quantities of rice plants, weeds, and diversity of the arthropod community.

9.1   No-Tillage Rice Production

9.1.1   Ecological Services of Paddy Fields

The paddy field in Japan is 24,180 km^2 (MAFF 2017) in area, which is only 6.4% of the total land area. However, this area occupies 96.7% of the total wetland area and is an extremely important part in the ecological composition of the country. More significantly, the paddy field of our country is primarily used as an agricultural system to produce rice, which is the main staple food in Japan. However, results of research conducted based on various viewpoints in recent years point out that paddy fields have various other functions in addition to producing rice (Matsuno et al. 2006; Natuhara 2013). The paddy field plays an important role in the various ecosystem functions such as flood control, groundwater recharge, improvement of water quality, local climate mitigation, fish culture and other non-rice productions, fostering culture and landscape, and maintaining biodiversity, among others. For example, the total water storage capacity of paddy fields is estimated to be 4 to 5 × 10^6 m^3 (NRIAE 1998) that is significant for flood control. Further, that paddy fields have the purifying capability of removing nitrogen and phosphoric acid from irrigation water is a well-known fact (Tabuchi 1998; Shiratani et al. 2005; Maruyama et al. 2008). Regarding the mitigation effect on local weather, the vegetation and water surface being known to reduce local temperature (Oke 1987) apply to paddy fields as well.

With respect to biodiversity, paddy fields are considered to be the source of various ecological functions. They provide an important habitat to foster biodiversity; Kiritani (2010) reports that the number of species inhabiting paddy fields exceeds 5000 based on literature information. Numerous reports indicate that the ecological functions of the paddy fields are declining rapidly due to urbanization, modernization, and abandonment of cultivation (Natuhara 2013). In other words, the remaining rice paddies in urban regions have an important role to alleviate the deterioration of the urban environment. For example, as the local climate mitigation effect is concerned, the daytime temperature during summer in regions with mixed paddy fields and urban areas has a maximum temperature reduction of about 2 °C as compared with the temperature in urban-only regions. In addition, a greater maximum temperature reduction is observed in regions with higher portions of paddy fields. The temperature reduction effect by paddy fields is significant in the region of about 150–200 m outward from the rim of the paddy field (Yokohari et al. 1997). Results of studies on the relationship between biodiversity and urbanization reported by Marzluff (2001) show that the species richness of birds decreases with urbanization. Additionally, according to McKinney (2008), the richness of species in mammals, amphibians, arthropods, and plants continues to decline as extreme urbanization progresses. In line with this, Nishida and Senga (2004) pointed out that small environments and networks of channels and levees are beneficial for organisms to exist in paddy fields located in the suburban regions of Japan.

Besides, the effect of flood control and the provision of spiritual comfort by paddy fields to urban residents are also considered important beneficial functions. Therefore, conserving paddy fields, regardless of how small they are, in the urban environment is considered necessary. Thus, paddy fields remaining in urban regions are expected to have various functions to alleviate environmental deterioration caused by urbanization. Moreover, in order to fully demonstrate these functions, it is necessary to pay attention to cultivation methods of paddy rice so that harmful effects of modernization of cultivation management are minimized. In particular, from the viewpoint of conserving biodiversity, reduction of chemical pesticide usage is of great importance. In urban regions, many residents live next to paddy fields; therefore, farmers in urban areas need to be extremely careful about the use of pesticides when cultivating rice in the paddy fields. Furthermore, many consumers living in urban regions demand safe or organic agricultural rice and other agricultural products that are produced with reduced or no pesticide in addition to seeking a comfortable space in the urban environment. Therefore, if farmers reduce the amount of agricultural chemical usage, the rice will have better market value and more business opportunity.

This chapter summarizes the measures, especially the conservation biological control methods, for reducing the amount of agricultural chemical usage in rice cultivation. In addition, the possibility of reducing pesticide usage on no-tillage cultivation of paddy rice is also discussed.

9.1.2 No-Tillage Cultivation of Rice Plants

Since the 1960s of the last century after World War II, conditions of rice cultivation in Japan have undergone a major transformation along with modernization of technology. From the end of World War II to the 1960s, productivity improved markedly due to variety change, early cultivation time, multiple application of chemical fertilizer and chemical pesticide, dense planting, etc. As a result, the self-sufficiency rate of rice reached 100% in the mid-1960s. However, during the latter half of the 1960s, the demand of rice decreased so that rice became overproduced in Japan. Therefore, from the 1970s onward, higher-quality and less labor-intensive production of rice was practiced, and the rice variety shifted to those with better taste. In addition, due to the progress of mechanization in transplanting and harvesting rice, as well as applying chemical fertilizers and pesticides, further saving in labor was achieved. But the cost of machinery, materials, and energy has increased. Meanwhile, farmers are facing the social and economic circumstances such as excessive production due to declining rice demand, relaxed import of cheaper foreign rice, as well as aging of farmers and extremely severe lack of successors. Therefore, further reduction of the production cost by reducing labor force requirement and input materials is of great importance. Meanwhile, there are strong concerns about both the safety and health of consumers of the agricultural products and environmental quality. Therefore, reducing the use of pesticides and chemical fertilizers is required from the viewpoint of protecting consumer health and environmental quality.

Farmers attempt to practice various cultivation methods such as direct seed cultivation and organic farming, among others. Along with these various cultivation methods, farmers also tackle the no-tillage practice to cultivate rice paddy without plowing and paddling at all. Plowing rice paddy during the fallow period of rice plants is carried out to promote the expression of soil nitrogen during the subsequent cultivation period (soil drying effect) and suppress the growth of winter/spring weeds. The paddling is done for the purpose of facilitating transplantation of seedlings, enhancing transplantation precision, promoting seedling initial growth, preventing water leakage, and controlling weeds. Both operations have extreme significance in cultivation control of paddy rice. Especially, paddling has been considered an essential process in rice transplanting cultivation. On the other hand, however, paddling is not always beneficial because the practice destructs the aggregate structure of the soil, inhibits the growth of rice root due to accumulation of harmful substances accompanying promotion of soil reduction, and impedes the harvesting work because of decreasing ground tolerance (Kumano et al. 1985). Also, the experience accumulated over the past 40–50 years shows that the working hours for carrying out the plowing/paddling work in cultivation control of paddy rice are about 10%–13% of the total work time during this 40–50-year period.

Studies on no-tillage cultivation have been carried out so far to address the issues of paddling. In particular, studies aiming at establishing practical cultivation methods for saving labor lead to the development and advancement of relevant agricultural machines. The Agriculture Research Center of Okayama Prefecture, Japan, developed a sowing machine for doing no-tillage direct seeding cultivation. In 1994, the no-tillage direct seeding cultivation of rice was carried out in the paddy field of about 100 ha (Okatake 1995). Moreover, during the latter half of the 1980s, machines for no-tillage transplanting cultivation of rice were improved, and farmers also began to tackle this practical technique (Kaneta 1994).

In parallel with the development of these sowing and rice transplanting machines, comparative studies on the relationship between the tillage/paddling method and the physicochemical properties of the soil have been conducted from the viewpoint of soil fertilizer science. The results indicate that cracks of the previous cultivation and the remaining pores formed by decomposition of rice plants in the no-tillage soil increase the vertical soil water permeability (Sato 1992). Thus, the soil of the no-tillage paddy field becomes oxidative (Sakai et al. 1968; Naganoma et al. 1989) and experiences less mineralization of soil nitrogen than the tillage soil. But the nitrogen expression timing is delayed, and the availability of nitrogen becomes maximum from the middle stage of growth of rice to the latter stage (Omori et al. 1970; Nonoyama and Yoshizawa 1976).

In addition, the increasing soil hardness in no-tillage soils makes the harvesting work easy (Naganoma et al. 1989). In no-tillage paddy fields, plant residues such as rice straw are not mixed with soil so that the amount of methane generated is smaller than that of conventional paddies because the soil is oxidative (Kaneta et al. 1992). With no-tillage paddy, there is no risk of river pollution because the falling water process does not occur after the paddling process (Hasegawa 1995). In this way, some knowledge is also obtained concerning the alleviated environmental load reduction effect of the no-tillage cultivation. Furthermore, the continuous no-tillage cultivation practice causes changes of annual weeds to perennial weeds. Therefore, it is necessary to change the type of herbicide applied for weed control (Arihara, personal communication) and control winter annual weeds before transplanting of rice seedlings. Hasegawa (1995) pointed out that more frequent applications of herbicides need to be practiced for the perennial weeds.

9.1.3 Natural Regulation of Insect Pests in No-Tillage Upland Crop Fields and Possibility of Natural Regulation of Insect Pests in No-Tillage Rice Paddy Fields

As mentioned in previous sections, no-tillage cultivation of rice was made practical by using sowing machines and transplanting machines as demonstrated in numerous cultivation studies. This technology has been widely accepted as a practical method to replace the tillage method. However, available fragmentary information has not provided convincing evidence to prove whether the environmental impacts of no-tillage cultivation of rice are positive or negative. Furthermore, very few case studies have been carried to examine the influence of no-tillage cultivation of rice on organisms inhabiting the paddy fields.

Soil erosion in upland fields became serious in the 1930s in the United States, and a movement to review the plowing method began with the objective of conserving the cultivated soil (Bennett 1935). In the early 1950s, herbicides were introduced, and the need for weed control by tillage diminished. For this reason, crop cultivation by using tillage reduction or no tillage was adopted by farmers and spread as "conservation tillage agriculture" (Stinner and House 1990). From the viewpoint of controlling pests, plowing in upland crops has the effect of controlling the pests surviving in the soil surface by reducing residues of the previous cultivation, thus destroying the pest habitats (Gebhardt et al. 1985). Stinner and House (1990) show that the no-tillage or reduced-tillage practice increases the population density of crop-damaging pests for 28% of the 51 species studied. Especially, soil-inhabiting pests have a tendency to increase more under reduced-tillage conditions. Moreover, 29% of the species were not affected, whereas 43% showed decreased population density or pest damage to crops. Hammond (1987) summarized that the conservation tillage did not result in an overall increase in pest species, but a change in the population and economic importance of individual species. Thus, in upland field crops, differences in pest species of concern in tillage, no-tillage, or reduced-tillage cultivation are observed.

Furthermore, according to Knipling (1992), tillage brings about an ecological imbalance in cultivated soil due to physical destruction of food resources and habitat as well as chemical pesticide application. For this reason, the tillage practice has a serious effect on not only insect pests but also beneficial insects such as the natural enemies of pests. In plot with no-tillage or reduced-tillage soil, residues of crops and weeds are deposited on the soil surface, and the extreme changes in soil temperature and humidity are alleviated (Griffith et al. 1986), thus providing a more stable habitat for arthropods living in the soil (Stinner and House 1990). In addition, predators such as spiders and carabid beetles often feed on decomposers such as Collembola and mites that inhabit the litter layer of the soil surface. For this reason, spiders and carabid beetles prefer no-tillage soils in which the litter accumulate to the soil surface where abundant decomposers exist (Stinner and House 1990; Wardle 1995). Actually, in various types of crops, the density of soil-inhabiting predators such as spiders and carabid beetles is known to be higher in no-tillage or reduced-tillage cultivation field (House and All 1981; Blumberg and Crossley 1983; House and Parmelee 1985; Paoletti 1987; Stinner et al. 1988; Kendall et al. 1991). In addition, Brust et al. (1985, 1986) conducted experiments to remove predators such as carabid beetles, wolf spiders, ants, etc. in corn cultivation fields with tillage or no tillage. The results showed that the predator density and the predation rate of the common cutworm (*Agrotis segetum*) are higher in the no-tillage field. In addition, Landis et al. (1987) showed that predation on pre-pupae of the oriental tobacco budworm (*Helicoverpa assulta* Guenée) by ants frequently occurs in corn no-tillage cultivation field compared with the tillage cultivation field. This observation suggests that the predator's impact on pests may be enhanced in the no-tillage/reduced-tillage cultivation field as described above.

In recent years, habitat management attempts to control the pest density by modifying the environment of the agricultural land ecosystem to promote natural enemies of the pests and their activities. This method of pest control by using natural enemies has received much attention, and the no-tillage or reduced-tillage cultivation is also regarded as an effective method of habitat management (Landis et al. 2000). As described above, either tillage or no-tillage practice in upland crop fields may directly modify the pest habitat and indirectly influence the pest predators and thus has a great influence on the population and dynamics of the pests. On the other hand, in conventional paddy rice cultivation, plowing, flooding, and paddling are carried out before rice transplanting. These procedures are thought to have a very great effect on the organisms inhabiting paddy fields, and the procedures seem to have even greater influence than tillage in upland crop fields.

Many animals use paddy fields as habitats during the fallow season. When the habitat is destroyed during flooding/paddling, they will either die or move out of the paddy field. Therefore, the animal community is established in paddy fields after transplanting by immigrating species. Although no tillage or no paddling is carried out in the no-tillage paddy fields, many animals are still forced to move out of the paddy field due to flooding. However, some residual rice and weed stubble that remain on the surface of paddy field become a temporary habitat for some animals. Hidaka (1993, 1997) cultivated Chinese milk vetch during the fallow period, and subsequently he tried to cultivate paddy rice in the same plot without tillage and paddling. As a result, the population density of lycosid spiders is high, whereas white-backed planthopper (*Sogatella furcifera* (Horváth 1899)) has low population density in the Chinese milk vetch herbage growing in no-tillage paddy fields as compared with conventional paddy fields. The authors' investigation also showed a tendency of higher population density for wolf spiders in no-tillage paddy fields located in Ibaraki Prefecture, Tochigi Prefecture, and Fukushima Prefecture than that in conventional paddy fields (Motobayashi, unpublished data). Spiders are known to play an important role as powerful natural enemies of rice insect pests (Itô et al. 1962; Heong et al. 1992; Way and Heong 1994). Kiritani et al. (1970) analyzed the life table of green rice leafhopper (*Nephotettix cincticeps*) and showed that predation by spiders is an important cause of death of the low-density leafhopper. In addition, Kiritani et al. (1972) estimated the quantity of prey captured by spiders in paddy fields by using the direct observation method. The results indicate that *Pardosa pseudoannulata* and *Ummeliata insecticeps* play important roles to control the population of the green rice leafhopper in paddy fields.

However, these results were obtained in the conventional paddy fields. Riechert and Lockley (1984) found that controlling pests only with spiders in many fields is difficult. They speculated that the decrease in spider density is caused by agricultural practices such as tillage, pesticide application, planting of crops, and harvesting, among others. The relationship between lycosid spiders and insect pest kinetics in no-tillage paddy fields, in which the density of lycosid spiders is expected to increase than in conventional paddy fields, has not yet been adequately studied. In addition, there has been few studies on spiders and natural enemies other than lycosid spiders.

9.1.4 Spider Assemblage in No-Tillage Rice Paddy Fields

To investigate the relationship between spiders and pests in no-tillage paddy fields, the author and colleagues set up experimental no-tillage and tillage plots in paddy fields located in Field Science Center of Tokyo University of Agriculture and Technology, Fuchu, Tokyo, to carry out the study (Motobayashi et al. 2006). The experiment was conducted between July and September for 3 years from 1999 to 2001. Numbers of spiders and other arthropods in each plot were counted by using the cylinder method (Southwood 1978). A cage with dimensions of $0.3 \times 0.2 \times 1.2$ m was placed on the rice plant; it used a suction tube to catch arthropods including those living on the surface of the water. Arthropod samples were collected from ten plants of each plot for 10 to 14 days. The spiders collected were immersed in 70% ethanol solution and identified; the number of individuals was counted for each species, and the spiders were categorized into each family. The samples were then dried at 40 °C for 48 h; their weights were measured with an electronic balance. For the period of 3 years, 6829 individual spiders (3993 from the no-tillage plots and 2836 individuals from the paddling plots) in seven families were collected (Table 9.1). Lycosidae was the most abundant family, followed by Tetragnathidae,

Table 9.1 Number of spiders collected in no-tillage and conventional tillage paddy fields in 1999–2001

Family	Species	Number	
		No tillage	Conventional tillage
Lycosidae		2456 (60.9[b])	1322 (46.1)
	Pardosa pseudoannulata (Bös et Str., 1906)	1811	942
	Pirata subpiraticus (Bös et Str., 1906)	642	376
	Juveniles of Lycosidae	3	4
Tetragnathidae		769 (20.0)	892 (32.4)
	Tetragnatha caudicula (Karsch, 1879)	104	85
	Tetragnatha vermiformis (Emerton, 1884)	23	9
	Tetragnatha maxillosa (Thorell, 1895)	16	6
	Tetragnatha squamata (Karsch, 1879)	14	7
	Pachygnatha quadrimaculata (Bös et Str., 1906)	129	183
	Pachygnatha clercki (Sundevall, 1823)	3	5
	Tetragnatha spp.[a]	240	272
	Dyschiriognatha spp.[a]	80	137
	Juveniles of Tetragnathidae	160	188
Linyphiidae		434 (10.8)	479 (16.6)
	Ummeliata insecticeps (Bös et Str., 1906)	20	35
	Gnathonarium exsiccatum (Bös et Str., 1906)	236	273
	Erigone prominens (Bös et Str., 1906)	5	2
	Juveniles of Linyphiidae	173	169
Salticidae		258 (6.4)	64 (2.2)
	Mendoza canestrinii (Ninni in Canestrini and Pavesi, 1868)	37	7
	Juveniles of Salticidae	221	57
Clubionidae		26 (0.6)	25 (0.9)
	Clubiona japonicola (Bös et Str., 1906)	3	2
	Juveniles of Clubionidae	23	23
Thomisidae		5 (0.1)	2 (0.1)
	Misumenops tricuspidatus (Fabricius, 1775)	3	0
	Juveniles of Thomisidae	2	2
Theridiidae		9 (0.2)	13 (0.4)
	Coleosoma octomaculatum (Bös et Str., 1906)	1	1
	Juveniles of Theridiidae	8	12
Unidentified		36 (0.9)	39 (1.3)
Total number		3993	2836

Adapted from Motobayashi et al. (2006)

[a]Juveniles that were not identifiable to species

[b]Numbers in parentheses denote the percentages of each family collected in each tillage treatment

Linyphiidae, and Salticidae. These families accounted for more than 97% of the total captured spiders; those in the families of Clubionidae, Thomisidae, and Theridiidae were extremely rare. As the species are concerned, individual spiders belonging to *Pardosa pseudoannulata* (Bös. et Str.) and *Pirata subpiraticus* (Bös. et Str.) of the Lycosidae family were abundant especially in the no-tillage paddy plots. In addition, *Tetragnatha caudicula* (Karsch), *Pachygnatha quadrimaculata* (Bös. et Str.), *Gnathonarium exsiccatum* (Bös. et Str.), and *Mendoza canestrinii* (Ninni in Canestrini and Pavesi) were also abundant.

The temporal changes of the abundance of spiders in both experimental plots were similar for 3 years. Although there was no difference in the number of spiders between the no-tillage plots and the tillage plots from May to early August and from mid-August to September, the number of spiders in the no-tillage plots had a tendency of being more numerous than in the tillage plots. Moreover, the amount of spider biomass in the no-tillage plots was slightly larger than in the tillage plots during the early stage of cultivation. As a result, the amount of spider biomass in the no-tillage plots was about twice much as that in the tillage plots (Fig. 9.1).

As mentioned in above paragraphs, various spider species that inhabit in paddy fields attack many rice pests such as leafhoppers and lepidopteran insects (Itô et al. 1962; Kiritani et al. 1972; Heong et al. 1992; Way and Heong 1994). The result of field experiments indicates that the number of spiders living in the no-tillage plot and the quantity of their biomass increase. This result conforms to the experimental result obtained with the Chinese milk vetch herbage/no-tillage paddy fields by

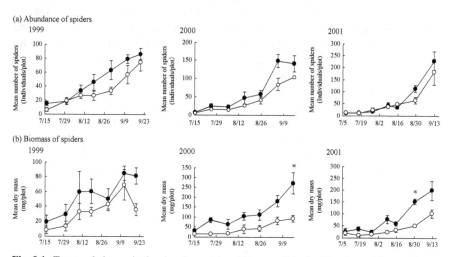

Fig. 9.1 Temporal change in the abundance (**a**) and biomass (**b**) of spiders in no-tillage and conventional tillage paddies. Closed symbols indicate no-tillage plots; open symbols indicate conventional tillage plots (Adapted from Motobayashi et al. 2006). Closed symbols indicate no-tillage plots, open symbols indicate conventional tillage plots, and vertical lines indicate standard error, *p < 0.05

Hidaka (1993, 1997). Further, results of other studies indicate that the population and biomass of generalist predators including spiders in no-tillage or reduced-tillage cultivation in upland fields increase (Brust et al. 1985; Brust and House 1990; Kendall et al. 1991; Symondson et al. 1996; Clark et al. 1997). The no-tillage or reduced-tillage cultivation system is shown to be a suitable environment for spiders to survive because of the complex habitat structure caused by increasing residues of crops and weeds, decreasing soil disturbance, rising soil moisture, or growing numbers of decomposers that inhabit in the soil and feed on the organic residues (Sunderland and Samu 2000; Wardle 1995). In conventional cultivation paddy fields, the habitats of spiders are destroyed by plowing, flooding, or paddling carried out before rice transplanting. In contrast, in the non-tillage cultivation paddy fields, some of the spiders including the lycosid spider or salticid spider may survive and multiply rapidly because the rice stubbles and weed residues are available for the spiders to live until after rice planting (Ishijima et al. 2004). In addition, this study further reveals that more dipteran insects live in no-tillage paddy fields than in the tillage paddies during the early stage of rice cultivation. Settle et al. (1996) and Murata (1995) suggest that dipteran insects are important prey for predators, including spiders inhabiting paddy fields. In the non-tillage paddy field, dipteran insects became a major food source for the lycosid spiders during the early cultivation period (Ishijima et al. 2006). Results of these observations show that insects in no-tillage paddy fields can be considered to play an important role for the growth of the lycosid spiders. These results also suggest that the complicated structure of soil and the abundant food supply available in the no-tillage paddy system promote the spider population (Fig. 9.2).

Incidentally, the compositions of the spider communities in no-tillage paddy fields and conventional tillage paddy fields are different. In other words, the number of individual spiders of the families Lycosidae and Salticidae in the no-tillage paddy fields is greater than that in the conventional tillage paddy fields, whereas the numbers of individual spiders of the families Tetragnathidae and Linyphiidae in both paddy fields are almost equivalent. This indicates that spiders of different families response to tillage practices differently. Kiritani (1986) and Hidaka (1998) proposed a hypothesis that stable habitats are advantageous for nonmigratory species, whereas unstable habitats are advantageous for migratory species. In winter wheat fields, nonmigratory spiders including *Pardosa* spp. are affected by management practices, but migratory spiders with ballooning including *Erigone* spp. are less affected by these management practices (Schmidt et al. 2005). For spiders inhabiting paddy fields, *P. pseudoannulata* and *P. subpiraticus* are considered nonmigratory species (Tanaka and Hamamura 1968; Kawahara et al. 1974), whereas *U. insecticeps*, *G. exsiccatum*, and *E. prominens* are migratory species. The migratory species overwinter in the paddy field during the fallow season and then move out in spring. They will migrate to paddy fields again in early summer by ballooning (Okuma 1974). Therefore, results of

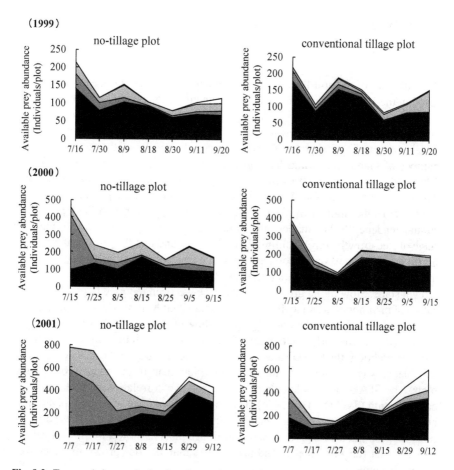

Fig. 9.2 Temporal changes in the abundance of potential prey of spiders in no-tillage and conventional tillage paddies in 1999–2001. (Adapted from Motobayashi et al. 2006) ■ : Homoptera, ▓ : Diptera, ▒ : Hemiptera, ☐ : Others.

this study are consistent with those reported by Schmidt et al. (2005). Furthermore, the response of the grand wandering coleopteran insects to the tillage practices has been shown to depend on the characteristics or life cycle of each species (Hance et al. 1990; Cárcamo et al. 1995; Clark et al. 1997). Therefore, the mechanism of spider community structure formation or increasing spider population in non-tillage paddy fields can be clarified by studying the migratory traits, response to prey animals, and life cycle, among others, for each species. For the migratory species, the environment around the paddy fields is also an important factor.

9.1.5 Natural Regulation of Insect Pests in No-Tillage Rice Paddy Fields

The influence of the no-tillage practice on the number of paddy rice pests is investigated in the following sections. The author and colleagues first compare the occurrence of the main pest species including the planthopper, leafhopper, rice leafroller, and straight swift in our experimental field during the rice-growing period.

Ishijima et al. (2004) have already reported the results of the survey about the number of planthoppers and leafhoppers conducted in our experimental field in 1999 and 2000. The pattern of number change for planthoppers and leafhoppers during this period was found to be similar to the pattern of the previous 2 years. In other words, the number of planthoppers and leafhoppers in no-tillage plots has a similar tendency to be smaller than in conventional tillage plots during the early stage of rice growth (July). Ishijima et al. (2004) found that the growth of rice in the no-tillage area is delayed as compared with the conventional tillage plots, thus causing a small number of planthoppers and leafhoppers to survive during the early growth stage. Slow initial growth of paddy rice has been reported to prevent colonization of planthoppers and leafhoppers (Hidaka 1993; Kajimura et al. 1993). In both years of 1990 and 1993, the no-tillage plots showed a slower increase in the number of stems during the early stage of growth as compared with the conventional tillage plots. In addition, the SPAD value, which indicates the degree of green color of the leaf blade, was also lower in the no-tillage plots than in the conventional tillage plots. The SPAD value is known to have a positive correlation with leaf nitrogen concentration (Tyubati et al. 1986). Thus, the nitrogen level in the rice plant of the no-tillage plots was lower than that of the conventional tillage plots. These observations suggest that the rice grown in the no-tillage plots has lower nutritional value in terms of diet for planthoppers and leafhoppers than the rice grown in the conventional tillage plots.

Furthermore, during the late growing season as mentioned in above sections, the number of individuals or the amount of biomass of lycosid spiders, which are considered to be effective natural enemies of the planthoppers and leafhoppers, was high in the no-tillage plots. In addition, Ishijima et al. (2006) reported that more than 60% of the diets for lycosid spiders were planthoppers and leafhoppers. Based on these facts, the predation pressure was exerted by the lycosid spiders against the planthoppers and leafhoppers in the no-tillage plots more strongly than in the conventional tillage plots. However, there is no difference between the numbers of planthoppers and leafhoppers in both treatment plots during the late growing season from 1999 to 2001. This is probably because the area of the experimented plot is small so that planthoppers and leafhoppers can move freely between the two plots.

The number of rice leafrollers (*Cnaphalocrocis medinalis* (Guenée, 1854)) tended to increase toward the later stage of cultivation (September). There was a trend that numbers of middle- and old-stage larvae of this species in the conventional tillage plots were larger than in the no-tillage plots. However, there was no

Table 9.2 Life table of *P. guttata guttata* in no-tillage and conventional tillage paddy plots in 2000

Developmental stage (x)	Mortality factor (dXF)	No-tillage plot			Conventional tillage plot		
		No.[a] surviving at start of stage interval (lx)	No.[a] dying within stage interval (dx)	% of mortality rate (100 qx)	No.[a] surviving at start of stage interval (lx)	No.[a] dying within stage interval (dx)	% of mortality rate (100 qx)
Egg		1130			1520		
	Parasitoid		210	18.6		270	17.8
	Unknown		70	6.2		70	4.6
			280	24.8		340	22.4
Instar I		850			1180		
	Failure in settlement		223	26.2		372	31.5
	Physiological death		15	1.8		40	3.4
			238	28.0		412	34.9
Instar II		612			768		
	Physiological death		0	0.0		0	0.0
	Unknown		17	2.8		48	6.3
			17	2.8		48	6.3
Instar III		595			720		
	Parasitoid		20	3.4		0	0.0
	Physiological death		35	5.9		20	2.8
	Unknown		35	5.9		70	9.7
			90	15.1		90	12.5
Instar IV		505			630		
	Parasitoid		50	9.9		70	11.1
	Physiological death		5	1.0		0	0.0
	Unknown		250	49.5		170	27.0
			305	60.4		240	38.1
Instar V		200			390		
	Parasitoid		70	35.0		110	28.2
	Physiological death		0	0.0		10	2.6
	Unknown		105	52.5		160	41.0
			175	87.5		280	71.8
Pupa	Parasitoid	25			120		
	Physiological death		20	80.0		80	66.7
			5	20.0		0	0.0
			25	100.0		80	66.7
Adult		0			40		
Accumulated mortality				100.0			97.4

Adapted from Motobayashi et al. (2007)
[a]Numbers are per 100 hills

difference in the numbers of larvae at young stage in both treatment plots, and thus, it was considered that some type of mortality factors affected middle- and old-stage larvae specifically in the no-tillage plots.

The number of larvae and pupae of the straight swift (*Parnara guttata guttata* (Bremer et Grey, 1852)) reached a peak value in early August and then declined gradually. The number of this species from the beginning of August to the beginning of September was slightly smaller in the no-tillage plots. This is because the number of older larvae and pupae in the no-tillage plots tended to be smaller than in the conventional tillage plots (Table 9.2). This was similar to the trend discussed for leafrollers in the aforementioned paragraphs.

As discussed in above sections, there was a difference between treated plots in the number of rice leafrollers and straight swifts. So we created a life table for the straight swift in each treated plot and compared the mortality process of this species. In addition, in order to quantify the impact of spiders on the straight swift larvae, an experiment on spider removal was conducted in both treatment plots.

In the life table of the straight swift created this time, parasitism is the most important factor contributing to the mortality of parasites during the egg and younger larval stages. The predatory parasitic species observed has already been

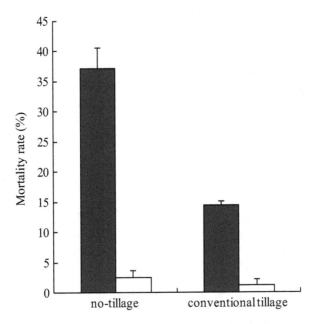

Fig. 9.3 Mortality rate of the inoculated fourth instar larvae of *P. guttata guttata* as a function of spider removal and tillage treatment. (Adapted from Motobayashi et al. 2007). ■, unmanipulated; □, spider removal; vertical lines indicate standard error

reported as a primary parasite of this species (Nakasuji 1982; Matsumura 1992). Additionally, both treatments show no significant difference in the mortality rate caused by these predatory parasites. Moreover, the unknown cause of death during the older larval stage is the main mortality factor, and this mortality rate is significantly higher in the no-tillage plots. As mentioned earlier, the number of spiders or the amount of biomass was higher in tillage plots than in conventional tillage plots from late August to early September. Predators other than spiders, such as wasps, birds, and frogs, have not been observed in this study. Furthermore, in the spider removal experiment, the mortality rate of the larvae in the spider non-removal plots was significantly higher than that in the spider removal plots, and the mortality rate of the larvae in the spider non-removal/no-tillage plots was higher than that in the spider non-removal/conventional tillage plots (Fig. 9.3) (Motobayashi et al. 2007). The results of such spider removal experiments and the information in the life table suggest that predation by spiders (especially lycosid spiders) is an important mortality factor for older larvae of the straight swift, and the effect is greater for the no-tillage paddy fields than the conventional paddy fields.

9.2 Rice Farming Supported by Crossbred Ducks

9.2.1 Rice-Duck Farming in Paddy Field

In western Japan, an integrated rice farming method that uses crossbred ducks, known as Aigamo farming in Japanese, has been practiced since the early Meiji era (150 years ago); this rice farming method is referred to as "rice-duck farming." In this method, there is no need to use agricultural chemicals for weed and insect control because the ducklings released and raised in the paddy fields eat the weeds and insects. The ducks that were initially used were the selectively bred imported from Europe and the United States. After World War II, the rice-duck farming practice declined owing to the propagation of agricultural chemicals and machines. Since around 1980, there has been high demand for rice with better quality and safety and a global demand for changes to conservation-oriented agriculture; therefore, attentions have been focused on rice-duck farming again. Further, after Furuno (1993) and Manda et al. (1993a) conducted scientific studies on the rice-duck farming method and demonstrated its effects, this method has spread nationwide.

The crossbred duck, which is a species of mallard, is a crossbreed of the wild mallard and poultry duck. It is also called a "call duck," because it has been used as a decoy for hunters to hunt ducks. Its meat is edible but has not been raised primarily for meat because this duck has slower growth than ordinary ducks. However, the economic effect of this breed is high because its slow growth roughly synchronizes with the rice plant growth so that simultaneous harvest of rice and production of duck meat can be achieved. Manda et al. (1993b) reported that crossbred ducks are nocturnal and actively forage from midnight to early morning. However, their predators such as wild dogs, weasels, and raccoons, among others, are also nocturnal. Hence, some preventive measures to protect crossbred ducks from their natural enemies at nighttime limit their intrinsic activities. In the author's previous study, ducklings were immediately attacked by crow after their release to the paddy fields. The idea of replacing ducklings with a small weeding robot has been studied by some researchers for saving operating and management costs (Chen et al. 2003).

The rice-duck farming is a promising technique for producing organic rice in Southeast Asia. Suh (2015) stated that the farming system involves organic food certification systems, organic farmers' cooperatives, community-wide organic farming, localized technical extension and educational services, and integration of farms and rice-duck. A comprehensive package of these institutional tools will further expedite the expansion of the integrated rice-duck farming particularly in low-income Southeast Asia, where the rice-growing or duck farming landscape is overwhelmingly dominated by smallholders. Li et al. (2012) pointed out that the rice-duck farming practice may have 20% lower yield than conventional rice cultivation because of damages caused by poorly controlled insects. The loss of rice yield for the farmer can be compensated by higher price of organic rice and the extra income from ducks; meanwhile, environmental issues due to use of chemicals and pesticides will be alleviated or even eliminated. The results may contribute to developing weed management systems in organic farming and enhance the transition from a herbicide-dependent cropping system to a more environmentally friendly cropping system.

9.2.2 Biological Diversity in Paddy Fields

The function of paddy fields is not limited to producing rice; it also includes flood prevention, soil erosion controls, preservation of air and water qualities, groundwater recharge, suppression of the heat island phenomenon, and townscape planning. More importantly, paddy fields nurture rich biodiversity owing to the presence of marshy ground. People have obtained the benefits from this ecological service by inheriting paddy rice farming successively for generations. Although many paddy fields have vanished in urban regions because of urbanization, some people still continue the activities to prepare and maintain paddy fields for these benefits. The Tokyo metropolitan government, based on "Ordinances and Enforcement Regulations Concerning Protection and Recovery of Nature in Tokyo," has been financially supporting groups such as residents' association to care for paddy fields remaining in urban regions (Oishi et al. 2012).

Paddy fields consist of a variety of ecosystems. Aquatic organisms such as rice-fish (*medaka* in Japanese), newts, and dytiscids live in the floodplains and back marshes of rivers, and natural swamplands can still be found in ponds and waterways around paddy fields. However, these organisms are at risk of extermination caused by activities such as draining paddy field or invasion by alien species. Tani et al. (2016) investigated the effects of changes of frogs inhabiting in waterways and farming lands in Sakai City. The construction of city-planned roads causes the paddy fields to fractionalize that affects the ecological network of paddy fields. With the construction of the public sewerage system, the water quality in the ponds and waterways is improved; however, the decreased volume of incoming water into dammed water bodies and ponds adversely influences the ecosystem. Some species such as the black-spotted pond frog, which is highly dependent on the waterways and watersides, are seriously impacted by this increased dryness caused by urbanization. Conversely, well-draining paddy fields favor some other species such as Japanese tree frogs, which pass the winter on land, as evidenced by many individuals present in paddy fields.

Fish living in the water of paddy fields is an important index for understanding the biological diversity. Mori (2017) reported that weather loaches use paddy fields as a site for reproduction and growth; they move through agricultural waterways including ponds to pass the winter in permanent water bodies. Reconstruction of canals and improvement of paddy fields degrade the waterways and ecology. This is because the water network interconnections such as the paddy field-waterway and the waterway-paddy field cross-connections are reduced so that suitable places for aquatic creatures to eat, spawn, and grow are decreasing. Therefore, maintaining such networks contributes to the preservation of biological diversity.

Hamada et al. (2015) investigated the habitats of birds by focusing on the continuity of paddy fields. They used GIS to set a 10 m × 10 m area as a cell and calculated the continuity of paddy fields and farming land of the adjacent cells based on land utilization. The observed birds were categorized according to their appearance in waterfront, forests, grassland, and urban as well as populations in various categories for comparisons. In the paddy fields with high continuity, waterfront birds such as *Egretta garzetta* Linnaeus and *Streptopelia orientalis* Latham are numerous, and there is a marked difference of the bird population during the breeding season. In the region with a high-paddy field area, *Motacilla grandis* Sharpe and *M. alba lugens* Linnaeus were often observed during the winter season. Conversely, many urban birds such as *Passer montanus* Linnaeus and *Corvus corone* Linnaeus were found in the area containing a high percentage of paddy field, and the high continuity of paddy fields was not relevant. The higher continuity of land use tended to feature more specific species; therefore, mosaic-like spaces such as the urban area contribute to biological diversity.

9.2.3 Efforts for the Restoration of the Ecology of Paddy Fields

Based on a study carried out by Koganezawa (2009), 14 agricultural cooperative associations have been working together on rice farming by reducing the use of pesticides and chemical fertilizers in Miyagi Prefecture. The rice cultivated using this method has been certified as "specially cultivated rice" by the prefecture. To maintain the entire environment of the region and enrich the ecological service, a distribution system (the price system) to help farmers continue agricultural activities in conjunction with environmental conservation is required. In this prefecture, farmers benefit from selling the high-priced branded rice as "environmental conservation rice."

Recently, there has been an attempt to flood paddy fields after rice harvest to maintain a proper water level to sustain the ecology of the fields throughout the entire year. The area known as Yatsu in Japanese is a rural natural marsh where paddy fields are filled with the water seeping out to the village for the year. The objective of flooding paddy fields after harvest was to create a similar environment for the maintenance of biota. Okada et al. (2016) investigated the ecology of nematodes in the paddy fields flooded year-round and showed that the very few number of nematodes inhabiting in the top 0–50 mm of the flooded paddy fields is less than half of that living in the ordinary paddy fields. The species of nematodes found in the flooded fields are about one-third of those existing in the ordinary paddy fields. This observation might be related to the suppressed growth of fungi and the slow growth of rice plants in the year-round flooded paddy fields as well. Wakasugi and Fujimori (2005) studied the ecology of dragonflies in year-round flooded fallow fields and reported that larvae living in the year-round paddy fields could endure the winter with a high survival rate. Further, the year-round flooded paddy fields are the wintering spots for migratory birds. These authors introduced some areas that progressed to create the branded rice by applying the excrement of migratory birds in no-till organic farming.

Table 9.3 Experimental condition of rice-duck farming

	Ten-duck	Five-duck	Control	Herbicide
Field area (m²)	250	250	200	200
Number of crossbred ducks	10	5	0	0
Herbicide	–	–	–	Modaun 1 kg

9.3 Simultaneous Production of Rice and Crossbred Ducks

9.3.1 Rice-Duck Farming Experiment at University Farm

9.3.1.1 Growth of Rice Plants and Crossbred Ducks

An experiment was undertaken for the simultaneous cultivation of rice plants and baby crossbred ducks from 1997 to 1999 at the farm of Tokyo University of Agriculture and Technology in Fuchu, Tokyo. This experiment aimed at investigating the effects of the release of crossbred ducks on the growth and quantities of rice plants, weeds, and the arthropod community (Yoshizawa 2000; Tojo et al. 2007). The data collected since 1999 are summarized in the following paragraphs.

The 2-week-old "Aokubi Aigamo" that were hatched in an artificial hatchery in Chiba Prefecture were released to the paddy fields 1 week after transplanting rice seedlings on June 23 until the time when ears of grain had appeared for 2 months on August 19. As shown in Table 9.3, the experiment was carried out using two densities of duck release: the five-duck plot with five crossbred ducks and the ten-duck plot with ten crossbred ducks released to a 250 m² paddy field. The control plot includes a plot with neither release of crossbred ducks nor use of herbicides, and the herbicide plot includes the use of herbicide only without release of ducks. During the release period, the compound feed "Power Layer" available from the National Federation of Agricultural Cooperative Association (JA) was fed to the crossbred ducks every day. The feed schedule was 9 g per duck after duck release from June 23 to June 30, 18 g from July 1 to July 10, 27 g from July 11 to August 10, and 45 g from August 11 to August 19.

The rice cultivar used was "Hitomebore, Tohoku 143," which was seeded on May 19 and planted on June 16. The basal fertilizer was only compost 300 kg per 1000 m² for each plot. The planting density was 18 plants per square meter in the paddy fields that were divided into four plots using leak prevention sheets. The

Fig. 9.4 View of bower for crossbred ducks (Aigamo)

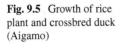

Fig. 9.5 Growth of rice plant and crossbred duck (Aigamo)

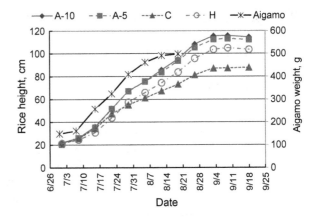

Table 9.4 Result of rice production in experimental plots

Plot	Ear per head	Ear per unit area (m⁻²)	Grain per ear	Ripening rate (%)	Weight of 1000 grains (g)	Yield per unit area (g m⁻²)
Ten-duck	20	418	79	71.1	21.9	536
Five-duck	21	381	79	72.8	21.5	455
Control	10	151	54	85.9	21.3	140
Herbicide	18	327	69	80.4	21.7	406

crossbred bird plots were surrounded by netting and covered with bird net to protect them from crows. A small shed for the ducks to rest was placed at the corner of the paddy fields (Fig. 9.4). In the herbicide plot, 1 kg of the herbicide "Modaun" was spread on June 23.

The growth of the crossbred ducks and the rice plants is shown in Fig. 9.5. On September 18, the height of the rice plants was 115 cm in the ten-duck plot, 112 cm in the five-duck plot, 103 cm in the herbicide plot, and 88 cm in the control plot. Therefore, the growth of the rice plants in the crossbred duck plots was faster than that in the control plot. The numbers of stems were 21 in the ten-duck plot, 19 in the five-duck plot, 16 in the herbicide plot, and 8 in the control plot; thus, more stems grew in the crossbred duck plots than in the herbicide and control plots. The leaf color was markedly light only in the control plot; however, it was similar in the other three plots. Light leaf color is caused by low nitrogen content, and the reason for low nitrogen content in the control plot was competition for nitrogen by weeds. Thus, the height of the rice plants and the number of stems in the crossbred duck plots are apparently better as compared with those in other experimental plots.

The rice yield for each experimental plot as shown in Table 9.4 was 536 kg in the ten-duck plot, 455 kg in the five-duck plot, 406 kg in the herbicide plot, and 140 kg in the control plot. There was a significant difference in the numbers of grains per ear for rice plants grown in different plots; the experimental plots with the crossbred ducks had the greatest numbers of grains per ear. However, the percentage of ripened grains tended to be lower accordingly. The author estimates that delaying the harvest term for 10 days, which leads to decreased protein content in the rice, improves the taste of the rice as well as lowers the quantity of immature rice and increases the yield.

Tiller emergence after harvest was different among the experimental plots. Tillers of rice grow in the case that the produced nourishment still stays in the plant root after ripening. In the plots with the crossbred ducks, there was no tiller right after harvest, whereas the tiller emergence occurred in the other plots that have no crossbred ducks. Therefore, the release of crossbred ducks increased the quantity of rice on the rice plants because the accumulated nourishment in the roots translocated to the ears without loss during the ripening period.

The crossbred ducks increase weights along with the growth of rice plants; however, the growth rate of ducks became slower during August. The weight increase during the sprouting season, the mean body weight of the crossbred ducks, and the standard deviation at this time were 499 ± 157 g. Because crossbred ducks with this body weight have no market value, they will be raised until their weights reach approximately 1 kg on average; they will then be sold to the meat suppliers.

9.3.1.2 Weed Situation

As shown in Table 9.5, weeds were not found in both crossbred duck plots, whereas they overgrew in the control plot and were found to exist in the water of the herbicide plot. A few Japanese millets coexisted in the two plots with the crossbred ducks. The herbicide suppressed the emergence of Japanese millets in the herbicide plot; however, *Monochoria vaginalis* was observed to overgrow seriously (Fig. 9.6). Overgrowth of both Japanese millet and *M. vaginalis* was seen in the control plot. The crossbred ducks had continuous suppression effects on *M. vaginalis* during the release period, whereas the herbicide had only transient suppression effects. These results are consistent with those reported by Manda et al. (1993a). The crossbred ducks were considered not to eat Gramineae; therefore, the Japanese millets could survive and grew along with the rice plants in the plot. Thus, the millet growth in the plot was not effectively suppressed or controlled in the crossbred duck plots.

Table 9.5 Weed emergence in experimental plots (unit: g DM)

Plot	July 18		Aug. 18		Sept. 18	
	Japanese millet	*Monochoria vaginalis*	Japanese millet	*Monochoria vaginalis*	Japanese millet	*Monochoria vaginalis*
Ten-duck	0.00	0.00	3.62	0.00	21.43	0.00
Five-duck	0.16	0.01	48.77	0.00	30.28	0.05
Control	4.12	9.33	53.27	89.70	73.27	46.78
Herbicide	0.07	0.46	0.24	36.50	0.00	28.96

(a) Ten-duck   (b) Five-duck

(c) Control   (d) Herbicide

Fig. 9.6 Views of water surface in each experimental plot. (**a**) Ten-duck. (**b**) Five-duck. (**c**) Control. (**d**) Herbicide

9.3.1.3   Appearance of Arthropods

As shown in Fig. 9.7, fewer arthropods were observed to exist in both experimental plots with the crossbred ducks. Overgrowth of weeds in the control plot caused the emergence of greater numbers of arthropods. The most observed arthropod during the research was the larva of the leafhopper (Deltocephalidae) that was numbered 25 per rice plant in the control plot, whereas fewer leafhoppers of an average 0.5 and 0.2 per rice plant were found to exist in the ten-duck and five-duck plots, respectively. Pyralidae (*Cnaphalocrocis medinalis* Guenée) eat the leaves of rice plants during the larval period (Fig. 9.8). When the larvae are normally covered by leaves, they are less likely to be eaten by crossbred ducks. Thus, the emergence of arachnids in the crossbred duck plots was about the same as in the other plots. Further, the number of arachnids tended to increase temporarily after the crossbred ducks were picked up. The reason was considered that after they were free from the threats of the crossbred ducks, the arachnids invaded the plots to seek for the insects as food.

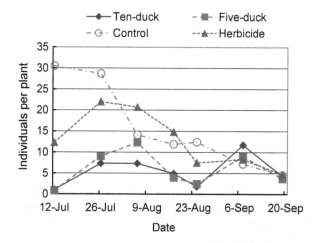

Fig. 9.7 Change in total individuals of arthropod in each experimental plot

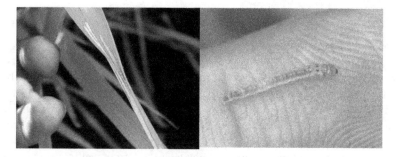

Fig. 9.8 Larva of *Cnaphalocrocis medinalis* and feeding damage on leaf

In conclusion, the experiment discussed in above paragraphs on simultaneous growing of crossbred ducks and cultivating of rice plants shows that the cultivation of the rice plant without using chemical fertilizers is possible.

9.3.1.4 Nitrogen Cycle

The cycle of nitrogen, which is an essential ingredient of fertilizers, is discussed in the following sections.

Input Amount of Nitrogen

The amount of nitrogen contained in the feed given to the crossbred ducks was calculated to be 3.01%. Multiplying this amount by 1494 g of feed consumed per duck during the release period, the total amount of nitrogen input per 1000 m² was

1.8 kg N in the ten-duck plot and 0.9 kg N in the five-duck plot. Therefore, the amount of nitrogen input contributed by duck feed is much smaller as compared with the conventional cultivation method. However, the nitrogen contained in the weeds and arthropods that were consumed by the ducks was eventually excreted back into the paddy fields.

Nitrogen Uptake of Rice Plants and Weeds

The total amount of nitrogen uptake per 1000 m² by both rice plants and weeds during the cultivation was 20.1 kg N in the ten-duck plot, 11.5 kg N in the five-duck plot, 11.3 kg N in the control plot, and 15.5 kg N in the herbicide plot (Fig. 9.9). In the control plot, the percentage of uptake by weeds was larger than that by the rice plants. The nitrogen contained in *M. vaginalis* was 1.17%, which was higher than the nitrogen contained in ears (0.94%), stems (0.66%), and roots (0.65%) of the rice plants. This suggests that when overgrowing, *M. vaginalis* could compete with rice plants.

Amount of Nitrogen Content in Soil

The differences of nitrogen contained in the soils of the experimental plots before ear emergence and after harvest were 13 mg N/g soil in the ten-duck plot, +15 mg N/g soil in the five-duck plot, −11 mg N/g soil in the control plot, and −37 mg N/g soil in the herbicide plot. Only the five-duck plot shows increasing nitrogen content primarily because the duck excretory nitrogen exceeded the nitrogen uptaken by rice plants. In the control plot, the rice plants did not grow well; thus, the nitrogen contained in the soil showed small decrease. The difference of nitrogen content per

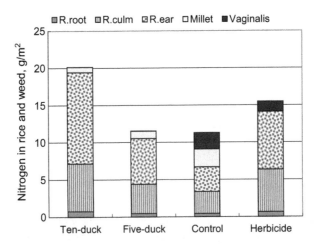

Fig. 9.9 Total nitrogen contained in rice and weed in each experimental plot

1000 m^2 of plot before the ear emergence and after the harvest in the ten-duck plot, the five-duck plot, the control plot, and the herbicide plot was −1.2, +1.4, −1.1, and −3.6 kg N, respectively. Thus, the cultivation of the rice plants in the plots with released crossbred ducks can alleviate the decrease of nitrogen contained in the soil.

Tashima et al. (2003) calculated the nitrogen balance of the paddy fields in which crossbred ducks and rice plants were cultivated simultaneously including nitrogen in the irrigation water. The results showed that the quantity of nitrogen provided by the irrigation water was similar to that from the crossbred duck excrement and that nitrogen contained in rice grain was nearly the same as the total nitrogen from crossbred duck excrement and irrigation water. Therefore, if only grains were harvested from the rice fields, the nitrogen cycle would be balanced in the paddy field. Weeds grow well during fallow in the paddy fields with crossbred ducks; thus, these authors suggested that the nitrogen nourishment could be gained from weeds plowed in the soil. Considering the nitrogen cycle over the entire cultivation cycle, the practice of simultaneous cultivation of the crossbred ducks and the rice plants should be implemented.

9.3.2 Challenges of the Simultaneous Rice-Duck Farming

The rice-duck farming may add additional costs, which are not associated with conventional rice production, to the rice farming practice. For crossbred ducks that must be purchased every year, newborn birds cost 300–500 yen each. When ducks are left in the field, ducklings or ducks may be attacked by wild dogs, weasels, and crows. Therefore, preventive measures such as installing nets or electric fences around the fields are necessary to protect the ducks. The costs for installing such equipment depend on the size of the fields. In addition, entrusting vendors for slaughtering the crossbred ducks also costs money; however, the duck meat can be sold at a high price. Additionally, the duck must grow to a certain size or weight to be marketed so that they should be fed continuously even after being removed from the paddy fields to ensure that they reach the marketable weight. The challenge to be faced by the management is to absorb the extra costs for such equipment and facilities. The rice so cultivated can be sold as "crossbred duck cultivated rice" at a high price. Some farmers can sell 60 kg of rice at a higher price of up to 2000–8000 yen more than the price of regular rice.

There is a high risk of pathogenic avian influenza caused by pathogenic viruses that will be brought over to the crossbred ducks in paddy fields. Gilbert et al. (2007) reported a clear and strong geographical correlation between the distributions of free grazing ducks and rice production. However, Henning et al. (2016) described that the potential contact between infected migrating birds through rice paddy fields and the crossbred ducks living in the same paddy fields plays an important role in spreading the avian influenza.

Utilizing crossbred ducks influences not only the weeding practice but the maintenance of the paddy field ecology. In western Japan, the breeding area of apple snails (*Pomacea canaliculata*) is spreading. Apple snails, also called the "giant river snails," damage the growth of rice plants and host parasitic worms in their bodies. Manda et al. (1993a) reported that released crossbred ducks suppressed the breeding of apple snails. However, adult apple snails are difficult to eat for ducklings so that crossbred ducks are assumed to prefer only juvenile ones. Thus, the simultaneous cultivation of crossbred ducks and rice plants is expected to eliminate the use of insecticides in the rice field.

Crossbred ducks, frogs, and carp are known to be released in paddy fields to cultivate rice plants. Teng et al. (2016) released 150 frogs weighing 20 g each in a 100 m² paddy field to examine the effect on the growth of rice plants. They reported that the growth of rice plants was not affected; but damages to rice plants caused by infectious and harmful insects were significantly decreased. In addition, numbers of bacteria and fungi in the soil were noted to increase. These authors stated that releasing frogs to paddy fields was an effective method to foster a protective artificial environment. Westwood et al. (2017) warned that the current use of chemical herbicides would increase the herbicide-resistant weeds so that shifting to the sustainable weed management becomes necessary. The future trend is to develop ecology-based technologies to achieve safe prevention and removal of weeds and maintenance of soil nutrients.

Global warming caused by emission of greenhouse gases from paddy fields is a serious issue in rice-producing countries. Sheng et al. (2018) reported that raising ducks in paddy fields can effectively reduce CH_4 emissions and thus significantly mitigate GWP (global warming potential) that is supported by the results of experiments carried out in Hubei Province, China. Analyses of GWP based on CH_4 and N_2O emissions demonstrate that compared with the traditional rice production, the rice-duck production significantly decreased the GWP by 28.1% in 2009 and 28.0% in 2010 and reduced the greenhouse gas intensity by 30.6% in 2009 and 29.8% in 2010. Zheng et al. (2017) suggested that the rice-fish-duck farming would build a more complex and diversified rice production system that reduces 15–30% water consumption as compared with the rice-fish or rice-duck system in addition to reducing methane emissions and thus protecting the environment. We should promote more research on developing methods to use animals for controlling pests and weeds with less consumptions of chemical fertilizer and pesticide.

References

Bennett AA (1935) Facing the erosion problem. Science 81:321–326

Blumberg AY, Crossley DA Jr (1983) Comparison of soil surface arthropod populations in conventional tillage and old field systems. Agro-Ecosystems 8:247–253

Brust GE, House GJ (1990) Effects of soil moisture, no-tillage and predators on southern corn rootworm (*Diabrotica undecimpunctata howardi*) survival in corn agroecosystems. Agric Ecosyst Environ 31:199–216

Brust GE, Stinner BR, McCartney DA (1985) Tillage and soil insecticide effects on predator-black cutworm (Lepidoptera: Noctuidae) interactions in corn agroecosystems. J Econ Entomol 78:1389–1392

Brust GE, Stinner BR, McCartney DA (1986) Predator activity and predation in corn agroecosystems. Environ Entomol 15:1017–1021

Cárcamo HA, Nienelä JK, Spence JR (1995) Farming and ground beetles: effects of agronomic practice on populations and community structure. Can Entomol 127:123–140

Chen B, Tojo S, Watanabe K (2003) Machine vision for a micro weeding robot in a paddy field. Biosyst Eng 85(4):393–404

Clark MS, Gage SH, Spence JR (1997) Habitat and management associated with common ground beetles (Coleoptera: Carabidae) in a Michigan agricultural landscape. Environ Entomol 26:519–527

Furuno T (1993) Aigamo Banzai. Rural Culture Association Japan, Tokyo (in Japanese)

Gebhardt MR, Daniel TC, Schweizer EE, Allmaras RR (1985) Conservation tillage. Science 230:625–630

Gilbert M, Xiao X, Chaitaweesub P, Kalpravidh W, Premashthira S, Boles S, Slingenbergh J (2007) Avian influenza, domestic ducks and rice agriculture in Thailand. Agric Ecosyst Environ 119(3–4):409–415

Griffith DR, Mannering JV, Box JE (1986) Soil and moisture management with reduced tillage. In: Sprague MA, Triplett GB (eds) No-tillage and surface-tillage agriculture. Wiley, New York, pp 19–57

Hamada A, Fukui W, Mizushima M (2015) The study of the relationship between connectivity of rural land use and urban fringe area, Kyoto city. J Jpn Soc Reveget Technol 41(1):145–150 (in Japanese)

Hammond RB (1987) Pest management in reduced tillage soybean cropping systems. In: House GJ, Stinner RE (eds) Arthropods in conservation tillage systems. ESA Misc. Publ. No.65. Entomol. Soc. Am., College Park, pp 19–28

Hance T, Gregoire-Wibo C, Lebrum P (1990) Agriculture and ground beetle populations. Pedobiologia 34:337–346

Hasegawa H (1995) Non-tilled rice cultivation. In: Koshihikari. Rural Culture Association, Tokyo, pp 448–454 (in Japanese)

Henning J, Pfeiffer DU, Stevenson M, Yulianto D, Priyono W, Meers J (2016) Who is spreading avian influenza in the moving duck flock farming network of Indonesia? PLoS One 11(3):e0152123

Heong KL, Aquino GB, Barrion AT (1992) Population dynamics of plant- leafhoppers and their natural enemies in rice ecosystems in the Philippines. Crop Prot 11:371–379

Hidaka K (1993) Farming systems for rice cultivation which promote the regulation of pest populations by natural enemies: planthopper management in traditional, intensive farming and LISA rice cultivation in Japan. FFTC Ext Bull 374:1–15

Hidaka K (1997) Community structure and regulatory mechanism of pest populations in rice paddies cultivated under intensive, traditionally organic and lower input organic farming in Japan. Biol Agric Hortic 15:35–49

Hidaka K (1998) Biodiversity conservation and environmentally regenerated farming system in rice paddy fields. Jpn J Ecol 48:167–178 (in Japanese with English abstract)

House GJ, All JN (1981) Carabid beetles in soybean agroecosystems. Environ Entomol 10:194–196

House GJ, Parmelee RW (1985) Comparison of soil arthropods and earthworms from conventional and no-tillage agroecosystems. Soil Tillage Res 5:351–360

Ishijima C, Motobayashi T, Nakai M, Kunimi Y (2004) Impacts of tillage practices on hoppers and predatory wolf spiders (Araneae: Lycosidae) in rice paddies. Appl Entomol Zool 39:155–162

Ishijima C, Taguchi A, Takagi M, Motobayashi T, Nakai M, Kunimi Y (2006) Observational evidence that the diet of wolf spiders (Araneae: Lycosidae) in paddies temporarily depends on dipterous insects. Appl Entomol Zool 41(2):195–200

Itô Y, Miyashita K, Sekiguchi K (1962) Studies on the predators of the rice crop insect pests, using the insecticidal check method. Jpn J Ecol 12:1–12

Kajimura T, Maeoka Y, Sudo T, Hidaka K, Nakasuji F, Nagai K (1993) Effects of organic farming of rice plants on population density of leafhoppers and planthoppers. Jpn J Appl Entomol Zool 37:137–144 (in Japanese with English abstract)

Kaneta Y, Awasaki Y, Yamaya S (1992) Soil management of heavy subsoil for the rotational use of paddy fields. 4 effects of straw application onto soil surface on production methane and growth of rice in the non tillage paddy fielfds. Tohoku Agric Res 45:77–78

Kaneta Y, Awasaki H, Murai Y (1994) The non-tillage rice culture by single application of fertilizer in a nursery box with controlled-release fertilizer. Jpn J Soil Sci Plant Nutr 65:385–391 (in Japanese with English abstract)

Kawahara S, Kiritani K, Kakiya N (1974) Population biology of Lycosa pseudoannulata (Bös. et Str.). Bull Kochi Prefect Inst Agric For Sci 6:7–22

Kendall DA, Chinn NE, Smith BD, Tidbodald C, Winstone L, Western NM (1991) Effects of straw disposal and tillage on spread of barley yellow dwarf virus in winter barley. Ann Appl Biol 119:359–364

Kiritani K (1986) Disturbance and re-stabilization of community. In: Kiritani K (ed) Japanese insect-Ecology of aggression and disturbance. Tokai University Press, Tokyo, pp 158–179

Kiritani K (2010) A comprehensive list of organisms associated with paddy ecosystems in Japan. The Institute of Agriculture and Natural Environment, Itoshima, 60pp

Kiritani K, Hokyo N, Sasaba T, Nakasuji F (1970) Studies on population dynamics of green rice leafhopper, Nephotettix cincticeps Uhler: regulatory mechanism of the population density. Res Popul Ecol 12:137–153

Kiritani K, Kawahara S, Sasaba T, Nakasuji F (1972) Quantitative evaluation of predation by spiders on the green rice leafhopper, Nephotettix cincticeps Uhler, by a sight-count method. Res Popul Ecol 13:187–200

Knipling EF (1992) Principles of insect parasitism analyzed from new perspectives. Practical implications for regulating insect populations by biological means. USDA, Washington, DC, p 337

Koganezawa T (2009) The relationship between rice production and ecosystem services: steps toward sustainable practices. Bull Miyagi Univ Educ 44:15–22

Kumano S, Seki K, Kon T (1985) Studies on puddling in mechanically transplanted rice cultivation. Bull Tohoku Natl Agric Exp Stn 72:1–53 (in Japanese with English abstract)

Landis DA, Bradley JR Jr, Gould F (1987) Behavior and survival of Heliothis zea prepupae in no-tillage and conventional-tillage corn. Environ Entomol 16:94–99

Landis DA, Wratten SD, Gurr GM (2000) Habitat management to conserve natural enemies of arthropod pests in agriculture. Annu Rev Entomol 45:175–201

Li S, Wei S, Zuo R, Wei J, Qiang S (2012) Changes in the weed seed bank over 9 consecutive years of rice–duck farming. Crop Prot 37:42–50

MAFF (2017) Summary of arable land area survey in 2017. http://www.maff.go.jp/j/tokei/kekka_gaiyou/sakumotu/menseki/h29/kouti/index.html

Manda M, Uchida H, Nakagama A, Matsumoto S, Shimoshikiryo K, Watanabe S (1993a) Effects of Aigamo ducks (Crossbred of wild and domestic ducks) herding on weeding and pest control in the paddy fields. Jpn J Poult Sci 30:365–370 (in Japanese)

Manda M, Uchida H, Nakagama A, Watanabe S (1993b) Growth and behavior of Aigamo ducks (crossbred of wild and domestic ducks) in paddy fields. J Poult Sci 30:383–387 (in Japanese)

Maruyama T, Hashimoto I, Murashima K, Takimoto H (2008) Evaluation of N and P mass balance in paddy rice culture along Kahokugata Lake, Japan, to assess potential lake pollution. Paddy Water Environ 6:355–362

Marzluff JM (2001) Worldwide urbanization and its effect on birds. In: Marzluff JM, Bowman R, Donelly R (eds) Avian ecology in an urbanizing world. Klawer, Boston, pp 19–47

Matsumura M (1992) Life table of the migrant skipper, Parnara guttata guttata Bremer et Grey in the northern peripheral area of its distribution. Appl Entomol Zool 27:331–340

Matsuno Y, Nakamura K, Masumoto T, Matsui H, Kato T, Sato Y (2006) Prospects for multifunctionality of paddy rice cultivation in Japan and other countries in monsoon Asia. Paddy Water Environ 4:189–197

McKinney ML (2008) Effects of urbanization on species richness: a review of plants and animals. In: Urban ecosystems. Springer, New York. https://doi.org/10.1007/s11252-007-0045-4

Mori A (2017) Biodiversity and conservation of fish living in paddy field ecosystem. J Rural Plann 35:482–487

Motobayashi T, Ishijima C, Takagi M, Murakami M, Taguchi A, Hidaka K, Kunimi Y (2006) Effects of tillage practices on spider assemblage in rice paddy fields. Appl Entomol Zool 41:371–381

Motobayashi T, Ishijima C, Murakami M, Takagi M, Taguchi A, Hidaka K, Kunimi Y (2007) Effect of spiders on inoculated populations of the migrant skipper *Parnara guttata guttata* Bremer et Grey in untilled and tilled paddy fields. Appl Entomol Zool 42:27–33

Murata K (1995) The interaction between spiders and prey insects under the sustainable cultivation. Acta Arachnol 44:83–96 (in Japanese with English abstract)

Naganoma H, Kaneta Y, Kodama T (1989) Soil management of heavy subsoil for the rotational use of paddy fields. I. Effect of partial rotary tilling and rice transplanting on soil condition and rice growth. Tohoku Agric Res 42:85–86

Nakasuji F (1982) Population dynamics of the migrant skipper butterfly Parnara guttata guttata (Lepidoptera: Hesperiidae) II. Survival rates of immature stages in paddy fields. Res Popul Ecol 24:157–173

Natuhara Y (2013) Ecosystem services by paddy fields as substitutes of natural wetlands in Japan. Ecol Eng 56:97–106

Nishida K, Senga Y (2004) Influence of environmental factors and paddy field on habitat of freshwater fishes at irrigation channel in an urbanizing area. Trans of JSIDRE 233:29–39 (in Japanese with English abstract)

Nonoyama Y, Yoshizawa T (1976) Studies on the soil management and fertilization method in the non-tilled direct seeding rice culture. 4. Influence of the mineralizing pattern of the soil nitrogen on the growth and yield of rice plant. Bull Chugoku Natl Agric Exp Stn E11:7–51

NRIAE (1998) The result of evaluation of public benefit accompanying agriculture and rural area by a replacement cost method. J Agric Econ 52:113–138 (in Japanese)

Oishi Y, Okubo K, Sasaki K (2012) A review and examination of effective measures for conservation of ecosystem in rice fields. J Fac Agric Shinshu Univ 48:15–22

Okada H, Niwa S, Hiroki M (2016) Nematode fauna of paddy field flooded all year round. Nematol Res 46:65–70

Okatake S (1995) Non-tilled direct seeding rice cultivation in Okayama prefecture. In: Challenge to direct seeding cultivation (vol 3) (Supervision by Kushibuchi K). Japan Association for Techno-innovation in Agriculture, Forestry and Fisheries Tokyo, pp 50–55 (in Japanese)

Oke TR (1987) Boundary layer climates, 2nd edn. Metheun, Lomdon, 435pp

Okuma C (1974) Aeronautic spiders caught by the trap net above paddy fields. Sci Bull Fac Agric Kyushu Univ 29(3):79–85

Omori T, Ono Y, Kawanaka K, Tsuboi I (1970) Soil chemical and nutritional studieson the ploughless planting by punch system sowing of rice and wheat plants in paddy field. I. Influence of the continued ploughless planting on the growth and yield of rice plants and physical and chemical properties of the soils. Spec Bull Okayama Prefect Agric Exp Stn 65:1–18

Paoletti MG (1987) Soil tillage, soil predator dynamics, control of cultivated lant pests. In: Stringanova BR (ed) Soil fauna and soil fertility. Nauka, Moscow, pp 417–422

Riechert SE, Lockley T (1984) Spiders as biological control agents. Annu Rev Entomol 29:299–320

Sakai H, Nonoyama Y, Komoto Y (1968) Studies on the soil management and fertilization method for the unplowing rice culture. I. High-yielding by fertilizer placement. Bull Chugoku Natl Agric Exp Stn E2:193–227

Sato T (1992) The land improvement of paddy field of heavy clay soil by no-tillage farming and prospect for multipurpose paddy field. J JSIDRE 60:723–728 (in Japanese)

Schmidt MH, Roschewitz I, Thies C, Tscharntke T (2005) Differential effects of landscape and management on diversity and density of ground-dwelling farmland spiders. J Appl Ecol 42:281–287

Settle WH, Ariawan H, Astuti ET, Cahyana W, Hakim AL, Hindayana D, Lestari AS (1996) Managing tropical rice pests through conservation of generalist natural enemies and alternative prey. Ecology 77:1975–1988

Sheng F, Cao CG, Li CF (2018) Integrated rice-duck farming decreases global warming potential and increases net ecosystem economic budget in Central China. Environ Sci Pollut Res 25:22744–22753

Shiratani E, Kubota T, Yoshinaga I, Hitomi T (2005) Effect of agriculture on nitrogen flow in the costal water environment at the Ariake Bay. Jpn Ecol Civil Eng 8:73–81

Southwood TRE (1978) Ecological methods. Chapman and Hall, London, 524pp

Stinner BJ, House GJ (1990) Arthropods and other invertebrates in conservation-tillage agriculture. Annu Rev Entomol 35:299–318

Stinner BR, McCartney DA, Van Doren DM Jr (1988) Soil and foliage arthropod communities in conventional, reduced and no-tollage corn systems: a comparison after 20 years of continuous cropping. Soil Tillage Res 11:147–158

Suh J (2015) An institutional and policy framework to foster integrated rice–duck farming in Asian developing countries. Int J Agric Sustain 13:294–307

Sunderland K, Samu F (2000) Effects of agricultural diversification on the abundance, distribution, and pest control potential of spiders: a review. Entomol Exp Appl 95:1–13

Symondson WOC, Glen DM, Wiltshire CW, Langdon CW, Liddell JE (1996) Effects of cultivation techniques and methods of straw disposal on predation by Pterostichus melanarius (Coleoptera: Carabidae) upon slugs (Gastropoda: Pulmonata) in an arable field. J Appl Ecol 33:741–753

Tabuchi T (1998) Science for clean water. Chapter V: Water cycle in watershed (Ed. Water Quality Environment Committee). Jpn Soc Irrigat Drain Reclamat Eng 100–107:115–118 (in Japanese)

Tanaka T, Hamamura T (1968) Population density of spiders in paddy field during winter. Bull College Agric Utsunomiya Univ 7:73–79

Tani M, Hara Y, Sampei Y (2016) Transformation of irrigated agricultural landscapes along with urbanization and its impact to distribution of frog species in Sakai city, Central Japan. Environ Inf Sci 30:237–242

Tashima F, Tatsumoto T, Egashira K (2003) Nitogen balance in the paddy field of the Aigamo-Paddy cultivation. Soil Sci Plant Nutr 74:5–21

Teng Q, Hu XF, Luo F, Cheng C, Ge X, Yang M, Liu L (2016) Influences of introducing frogs in the paddy fields on soil properties and rice growth. J Soils Sediment 16:51–61

Tojo S, Yoshizawa M, Motobayashi T, Watanabe K (2007) Effects of loosing Aigamo ducks on the growth of rice plants, weeds, and the number of arthropods in paddy fields. Weed Biol Manag 7:38–43

Tyubati T, Asano I, Oikawa I (1986) The diagnosis of nitrogen nutrition of rice plants (Sasanishiki) using chlorophyll-meter. Jpn J Soil Sci Plant Nutr 57:190–193 (in Japanese)

Wakasugi K, Fujimori S (2005) Influence that reformation into well-drained paddy field has on the habitation environment of dragonflies, and its measures. J JSIDRE 73:785–788

Wardle DA (1995) Impacts of disturbance on detritus food webs in agro-ecosystems of contrasting tillage and weed management practices. Adv Ecol Res 26:105–182

Way MJ, Heong KL (1994) The role of biodiversity in the dynamics and management of insect pests of tropical irrigated rice – a review. Bull Entomol Res 84:567–587

Westwood JH, Charudattan R, Duke SO, Fennimore SA, Marrone P, Slaughter DC, Swanton C, Zollinger R (2017) Weed management in 2050: perspectives on the future of weed science. Weed Sci 66:275–228

Yokohari M, Brown RD, Kato T, Moriyama H (1997) Effects of paddy fields on summertime air and surface temperatures in urban fringe areas of Tokyo. Jpn Landsc Urban Plan 38:1–11

Yoshizawa M (2000) Effects of releasing crossbreed ducks on the growth of rice plants, weeds, and the number of arthropods in paddy fields. BS thesis, Tokyo University of Agriculture and Technology (in Japanese)

Zheng H, Huang H, Chen C, Fu Z, Xu H, Tan S, She W, Liao X, Tang J (2017) Traditional symbiotic farming technology in China promotes the sustainability of a flooded rice production system. Sustain Sci 12:155–161

Chapter 10
Recycle-Based Organic Agriculture in Japan and the World

Yosei Oikawa, Vicheka Lorn, J. Indro Surono, Y. P. Sudaryanto,
Dian Askhabul Yamin, Tineke Mandang, and Seishu Tojo

Abstract In the first section, possibilities and challenges in project-type organic agriculture in Central Vietnam are examined. Findings and lessons learned from promoting organic agriculture in buffer zone villages of Bach Ma National Park are evaluated. The project team proposes that the multipurpose use of charcoal is an appropriate technology to be applied in this area. The objective is to improve their livelihood through the use of charcoal in both agricultural industry and livestock husbandry. Another project is creating organic vegetable supply chain in Hue City. This project aims to spread biogas digesters using livestock manure and organic vegetable cultivation using biogas residue.

In the second section, a corporate-type organic agriculture in Indonesia is introduced. This foundation has the cooperating farmers around the estate and cultivates more than 70 kinds of vegetable and horticultural crops all year round. They employ mechanical soil conservation techniques such as making bench terraces with beds following the soil contour or cutting slopes or perpendicular sides combined with vegetative techniques by planting grass or other plants to reinforce terraces. To disseminate this organic agriculture, the foundation regularly organizes organic farming training sessions and workshops, which are attended by various stakeholders such as students, farmers, practitioners, and other professionals to transfer knowledge from the experts to the participants.

In the third section, an emergent-type local organic agriculture in Japan is explained. A biogas plant established and operated by a nonprofitable organization

Y. Oikawa (✉) · S. Tojo
Institute of Agriculture, Tokyo University of Agriculture and Technology,
Fuchu, Tokyo, Japan
e-mail: yosei@cc.tuat.ac.jp; tojo@cc.tuat.ac.jp

V. Lorn
United Graduate School of Agricultural Science, Tokyo University of Agriculture
and Technology, Fuchu, Tokyo, Japan

J. I. Surono · Y. P. Sudaryanto · D. A. Yamin
Bina Sarana Bakti (BSB) Foundation, Bogor, Indonesia

T. Mandang
Bogor Agricultural University, Bogor, Indonesia

© Springer Nature Singapore Pte Ltd. 2020
S. Tojo (ed.), *Recycle Based Organic Agriculture in a City*,
https://doi.org/10.1007/978-981-32-9872-9_10

(NPO) is producing digested slurry and supplying it to the farming fields as organic fertilizer. The NPO provides detailed recipes for the use of digested slurry in the paddy and vegetable fields. Among the 397 commercial farmers in Ogawamachi, 33 engage in organic agriculture. These organic farmers are associated with the trainees working at a leading farm. After the training period, some trainees become new farmers and lead local groups to engage in organic farming.

10.1 Biochar Application for Rural-Urban Stakeholders: Case Studies in Central Vietnam

10.1.1 Biochar Application for Recycle-Based Agriculture in Tropical Developing Countries

Various urban and rural biomass can be utilized for designing and constructing a recycle-based agriculture in urban areas. In this section, possibilities and challenges of biochar application in urban recycle-based organic agriculture, particularly in developing countries, are examined. Major developing countries are usually located in tropical regions and have infertile and acidic soils. Use of biochar as a soil amendment is expected to improve such tropical soils. The issue of concern is what needs to be accomplished by applying biochar in urban recycle-based agriculture for tropical developing countries.

In the following paragraphs, findings of soil improvement using biochar are reviewed; and the lessons learned from promoting organic agriculture by using biochar in buffer zone villages of Bach Ma National Park, Vietnam, are evaluated as case studies of the recycle-based agricultural model in tropical rural areas. An example of recycle-based agriculture using biogas digester in suburban villages near Hue City that is about 40 km from Bach Ma National Park is also referenced. Based on these case studies and related information, possibilities and challenges of applying biochar in urban recycle-based organic agriculture in tropical developing countries are discussed.

10.1.1.1 Biochar as Soil Amendment

As mentioned in Chap. 3, application of biochar or charcoal has been studied in various fields such as agricultural engineering and environmental sciences, among many others, in recent years. Biochar studies cover a broad range of subjects from basic scientific research on materials, carbonization process, and physicochemical and biological properties to applied research on effects of using biochar or mixture of biochar and fertilizer to improve soil (Lehmann and Joseph 2009, 2015). The characteristics of biochar are capable of improving soil physical, chemical, and biological properties as a soil amendment (Steiner et al. 2007). It neutralizes acidic soils, enhances the availability of soil nutrients to plants, and improves the growth and yield of crops. In addition, the soil mixed with biochar increases the nutrient retention capacity, thus reducing the loss of soil nutrients and fertilizer (Glaser et al. 2002).

If farmers take advantage of these characteristics of biochar in organic agriculture, they will be able to develop the add-on values of organic farm products while improving the soil properties. This concept is confirmed by applying biochar in an organic farming system using locally available biomass such as livestock manure and crop residues to substitute chemical fertilizer. This combined application of biochar and organic fertilizer is expected to save chemical fertilizer and pesticide. Moreover, recycling of carbonizing agricultural wastes contributes to waste reduction and carbon fixation.

10.1.1.2   Bach Ma Charcoal Project

In the central region of Vietnam, arable land is limited, and the soil fertility is low. Additionally, this region is often devastated by typhoons and floods; thus, the land is not suitable for growing crops. Bach Ma National Park was established in 1991. It is located between Hue and Da Nang in Central Vietnam on the most seaside border of the forest corridor from Laos to the South China Sea (Fig. 10.1). Because some forest parcels in the buffer zone were transferred from the national park to the local communes through forest land allocation, acacia tree plantations and upland fields have been expanded to the area adjacent to the core zone of the protected forests.

The Bach Ma Charcoal Project was planned based on the idea that effective maintenance of core protected areas (core zones) will be achieved by improving the

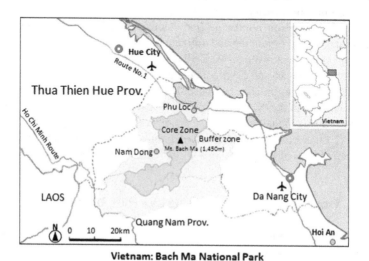

Fig. 10.1   Bach Ma National Park located between Hue and Da Nang in Central Vietnam

livelihood of the farmers living in the buffer zones surrounding the core zone. The deputy director of Bach Ma National Park, who completed a master's program at Tokyo University of Agriculture and Technology (TUAT) in 2006, proposed that the multipurpose use of charcoal, which he had learned in rural Japan, is an appropriate technology to be applied at this site (Yamada et al. 2014). Then, this project was considered as a grassroots technical cooperation with Tokyo University of Agriculture and Technology through the JICA (Japan International Cooperation Agency) Partnership Program for 5 years from 2008 to 2013. The title of this project was "Technical Cooperation Project for Improving Rural Living and Nature Conservation by Multipurpose Use of Charcoal and Wood Vinegar in the Bach Ma National Park." The objective is to improve the rural life in the buffer zone villages of the national park in Thua Thien Hue Province through the use of charcoal in both agricultural industry and livestock husbandry.

This charcoal project focused on transferring the charcoal technology to farmers, particularly targeting small-scale farmers who are interested in cultivating organic vegetables in their home garden. In this project, organic fertilizer containing locally available charcoal named as *Bokashi Than* that is made of rice husk charcoal (RHC), rice bran, and livestock manure is made available to farmers. The term *Bokashi Than* is coined by combining Japanese term *bokashi* and Vietnamese term *than*. Incidentally, the project consistently used the term "rice husk charcoal (RHC)," which will be used in this book as synonym of "biochar."

Prior to starting the charcoal project from July 2008 to June 2011, local residents used to gather firewood from acacia plantations, sell firewood in local markets, and buy necessary supplies for staples including vegetables. They planted perennial crops such as coconut palm, betel palm, mango, jackfruit, rambutan, and others besides upland crops such as corn, sweet potato, and cassava; but few farmers grew vegetables. These farming activities were practiced in Khe Su Village, Loc Tri Commune, Phu Loc District.

Our team worked on improving livelihoods of local people by implementing simple strategies: (1) producing and utilizing biochar from agricultural wastes in the areas around Bach Ma National Park, (2) breeding healthy livestock by using bio-char and wood vinegar, and (3) producing and selling safe and reliable agricultural crops produced using biochar compost. The series of these activities are summa-rized by our team in the "Charcoal-applied Environmentally friendly Farming with Livestock (CEFL) model" (Fig. 10.2). As a result, this farming model was trans-ferred to 12 other households in Khe Su Hamlet, which borders the national park; thus, dissemination of that model in a wider region became a follow-up task.

10.1.1.3 Organic Farming Model Using Rice Husk Biochar

The CEFL model is briefly outlined in Fig. 10.2. In this model, farmers first make rice husk biochar (RHB) and then feed the piglets with the mixture of a small amount of RHB powder and regular feed to prevent diarrhea, as well as spray a

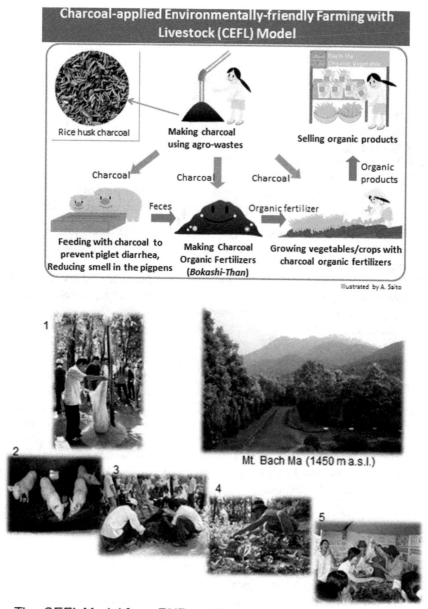

Fig. 10.2 CEFL model promoting organic vegetable farming using rice husk biochar

certain amount of RHB on the pigpen floor to reduce bad smell from the excreta. In addition, the ingredients of rice husk, RHB, and rice bran with livestock manure are mixed and composted at temperature of about 50–55 °C into organic fertilizer "charcoal Bokashi." Local farmers apply the compost for growing organic vegetables in their garden, and subsequently they sell the organic products on the market.

10.1.1.4   Outline of Dissemination Activities

The project to disseminate the CEFL model emphasized the following characteristics to promote farmers' active participation: "simple and easy to implement," "low cost by using agricultural waste," "reducing the use of pesticides and antibiotics," and "increasing agricultural income and self-sufficiency of food."

In order to sustain the technology transfer in more and wider regions, the following follow-up activities were upheld from September 2011 to March 2013 to establish a system for disseminating this model to residents throughout the buffer zone:

(1) Experts and local field assistants provided follow-up technical supports for the recycle-based agricultural model using biochar to the farmers participating in the training workshops.
(2) Teaching materials and training programs were prepared and revised through organizing training workshops. Experts in Vietnam and Japan collaborated to conduct field trials/experiments on biochar application and compiled a reference guidebook for the workshop participants to review after participating in training workshops.
(3) Training workshops were held for disseminating the farming model using biochar and for developing human resources to promote dissemination of the farming model. The training workshops were conducted six times for more than 150 participants during the follow-up period of 1.5 years. Thirty-six model farmers, who continued to practice actively even after the training workshop had been completed, were certified as model farmers and expected to disseminate the farming model to other farmers. In addition, field demonstrations and study tours were organized for exchanging technologies with organic farmers inside and outside the buffer zone.

10.1.1.5   Issues Observed on Collaboration Among Stakeholders

Several tasks have been clarified from participant observations after the project was ended. The CEFL model was originally targeted at the individual farms that could procure sufficient agricultural wastes. In practice, however, the collaboration among local stakeholders was necessary to realize and sustain the farming model. In Khe Su Hamlet, livestock manure, rice husk, and rice bran were scarce because the number of livestock and production of rice were limited. The farmers started vegetable cultivation using organic fertilizer *Bokashi Than* with RHB in a vegetable garden plot of 200 square meters for each home garden, and some members expanded the garden plot areas. The project initially needed to obtain these materials from other villages. The relations among vegetable growers, livestock-raising farmers, and rice farmers sustaining the biomass flows inside/outside the village are shown in Fig. 10.3.

In addition, although the charcoal project targeted small-scale farmers in the buffer zone villages, its small scale was incapable of coping with large demands of urban market. Hence, devising a framework such as farmers' market is necessary for small-scale farmers to sell their produce directly to consumers. However, from the viewpoint of securing customers, small-scale farmers did not have proper transportation to supply farm products to residents living in urban areas so that the farmers can only sell their produce to rural markets. Farmers in the buffer zone were far from urban consumption areas as compared with urban and suburban farmers. Therefore, they might need to count on local customers and the tourists visiting the national park as potential customers.

As shown in Fig. 10.3, Bach Ma National Park has been providing environmental education and ecotourism services to visitors, tourists, as well as rural and urban residents since before the charcoal project was implemented. Loggers and non-timber forest products (NTFP) collectors, who extracted forest resources in the protected forests, live in the buffer zone along with local residents. The national park has tried to change their activities by providing alternative income sources through eco-rural tourism such as village walk program as well as cultivation of medicinal plants and tropical fruits. Organic vegetable cultivation has the potential to be added as an alternative economic option for villagers in the buffer zone.

At the beginning of the project, vegetable-growing farmers in Khe Su Hamlet were unable to supply their products stably to local markets. Fresh vegetables, which are tasty (sweet and crispy), were often consumed by the growers themselves. Gradually, more model farmers increased their commercial production and

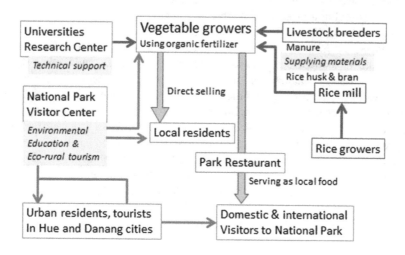

Stakeholders in Bach Ma Charcoal Project

Fig. 10.3 Stakeholders in the Bach Ma Charcoal Project

sold their harvests directly on the market. In the national park, cafeterias provided tourists and staff with meals prepared using the produce grown by farmers from Khe Su Hamlet. In 2012, the farmers attempted to supply organic agricultural products to a broker in Da Nang City for a short period of time, but they finally did not contract with brokers due to unsatisfied conditions. From this experience, the farmer group members of the charcoal project targeted local consumers near Cau Hai Market, located in the district center. They brought their harvest to the market and sold the produce at the booth.

Even small-scale vegetable producers may realize the benefit of supplying stable vegetable production when cooperating with other stakeholders who provide materials of organic fertilizer. This charcoal project demonstrates to the farmers a technical option to produce organic vegetables that contributes to improving livelihoods of the farmers. Development of organic agriculture using rice husk biochar was observed in the buffer zone and around Hue City (Oikawa et al. 2018). However, because the project primarily emphasized technology transfer, it was impossible to build a supply chain from rural producers in Bach Ma to urban consumers in Hue City and Da Nang City. A case study on developing the recycle-based organic agriculture near the urban market enables us to discuss the challenges of recycle-based organic agriculture using biochar in urban areas.

10.1.2 An Example of Recycle-Based Organic Agriculture in Hue City

In the following sections, we will examine the case of building a supply chain of organic agricultural products for urban consumers through recycle-based organic agriculture using biogas digester in the suburbs of Hue City.

Bridge Asia Japan (BAJ) is a nonprofitable organization (NPO) established in 1993 with headquarters in Tokyo. In Vietnam, Myanmar, and Japan, BAJ has engaged in international cooperation activities such as environmental conservation by providing technical opportunities, making available opportunities for capacity building, supporting revenue enhancement, and improving the infrastructure for regional development (BAJ Website n.d.). Since 2011, BAJ has introduced biogas digester to the farmers near Hue City in Central Vietnam and has been working on spreading organic vegetable cultivation using livestock manure and biogas residue. BAJ has been collaborating with the farmers to sell organic products at their farmers' market, toward a clear goal of improving livelihoods of farmers. In recent years, BAJ has been developing its own farm and working on increasing the awareness of organic agriculture to urban consumers.

10.1.2.1 Biogas Digester in the Integrated Farming System

Introduction of biogas equipment in Vietnam has started in the northern region from the 1960s and gained much advancement in the southern Mekong Delta since the 1980s (Nguyen 2011). In the Mekong Delta, an integrated farming system known as the VAC system consisting of fruit production (V = vườn), fish farming (A = ao), and livestock breeding (C = chuồng) has already become widespread. The VACB model that further includes the biogas digester in the VAC system has been proposed. In the VACB model, various economic and environmental benefits are expected. These benefits include (1) saving household expenses and family labor by using biogas as cooking fuel, (2) saving chemical fertilizer by using biogas digester effluent as an organic fertilizer, and (3) increasing crop yield by using biogas digester effluent. The environmental benefits include (1) avoiding smoke and soot when burning firewood, (2) reducing the smell of livestock excreta and of the population of flies and parasites, and (3) improving water quality of ponds and waterways (Yamada 2008).

In Thuy Xuan Ward of Hue City, located in a suburban area, many small-scale livestock farmers contaminated the surrounding environment due to direct discharge of untreated livestock excreta. The farmers need to improve their livelihood while solving the problems of the livestock excreta. Therefore, BAJ installed biogas digesters to process the livestock excreta derived from livestock farms in an appropriate manner for reducing environmental pollution in the surrounding areas and producing useful resources. At the same time, the installation of biogas digester was also expected to reduce greenhouse gas emissions by alleviating methane emissions and burning raw firewood and fossil fuels (BAJ 2012).

For installing the biogas digester system, BAJ formed a biogas committee with biogas digester installed farmers as committee members. The committee shared pertinent and useful information on the operation of biogas digester and effective use of the biogas and liquid fertilizer/sludge generated from biogas digester (BAJ 2013). In order to cover the operation and maintenance costs of the biogas digester system and capital costs for new facilities, biogas digester installed farmers participated in the reserve fund system. Thus, BAJ has built a model that is successful in developing the agricultural system and animal husbandry while conserving the environment. Thirty-six biogas digesters have been constructed through a 3-year grant project starting in 2011 (BAJ 2014).

10.1.2.2 Farmers Shop for Organic Agricultural Products

BAJ members aimed at not only transferring technology relevant to the construction and use of biogas facilities but also improving the farmers' livelihood further. They opened the first farmers' market named "Hue Farmers' Shop" that functions as a direct sales point in Hue City starting in December 2014 and started the second shop in August 2016 (Fig. 10.4). They have steadily increased the number of participating members and sales of organic agricultural products and have developed their direct-sale business (BAJ 2017). This created a supply chain of organic agricultural

Fig. 10.4 Hue Farmers' Shop of BAJ in Hue City

products by connecting producers and consumers and established a mechanism to keep the flow of biomass.

10.1.2.3 Contract Farming in Leased Land

In Hue City, the number of stores selling organic vegetables has been increasing rapidly in a few years (BAJ 2017). In order to differentiate their activities from competing stores, in April 2017, BAJ opened its own farm in a suburban area by leasing farmland directly from landowners. In the trial section of this farm, new crop varieties are introduced to increase the number of items sold at the farmers' market. In the farmers' exclusive section, participating farmers practice organic cultivation to improve their cultivation technology. Furthermore, an experiential section was set up for consumers (BAJ 2018).

Vegetables are intensively cultivated in Hue City and the suburbs, and some have been consumed by a large number of domestic and international tourists (Fig. 10.5). Due to the growing awareness of organic agricultural products and safety vegetables in recent years, local farmers will be able to increase agricultural products with high add-on values by not only implementing production technology but also developing marketing channels. In recent years, Que Lam Group has been developing nationwide production and sales of organic agricultural products and opening organic product shops and organic cafes in Hue City since 2017. Thus, opportunities for farmers to sell organic produce and for consumers to purchase the produce have been increasing not only in Hue City but also in major big cities such as Hanoi and Ho Chi Minh City.

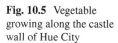

Fig. 10.5 Vegetable growing along the castle wall of Hue City

10.1.3   Possibility of Using Biochar in Organic Agriculture in Urban Areas

Organic farming in Bach Ma and the suburbs of Hue City is an example of using pig and cattle manure as organic fertilizer. To procure an appropriate quantity of live-stock manure, organic farmers may be required to collaborate with pig and poultry farmers. Additionally, they also need to pay attention to the quality of manure.

Another example model of practicing urban recycle-based organic agriculture in Central Vietnam is the organic cultivation by applying algae harvested from the lagoon to sandy soil. This practice is carried out in Tra Que Vegetable Village, which is on the outskirts of Hoi An Ancient Town, Quang Nam Province (Hoi An Tourism n.d.). Local farmers have been producing vegetables without using livestock and poultry manure.

While the number of tourists in Bach Ma National Park decreases significantly during the rainy season from October to January, in Hue and Hoi An, tourists continue to come throughout the year except during the period of flooding. Therefore, farmers are expected to supply organic vegetables for tourists throughout the whole year. If biomass is available in each area, practicing organic cultivation without using livestock manure may become possible, as practiced in Tra Que Vegetable Village.

In tropical regions, where organic matters are quickly decomposed and the local soil is acidic and infertile, applying biochar to the soil is expected to establish a recycle-based organic agriculture to effectively improve the soil and save fertilizer consumptions. Issues concerning this practice remain whether the local biomass is available for biochar and how to produce the biochar.

Rich bamboo vegetation is cultivated in Kameoka City, which is located in the west of Kyoto Prefecture, Japan. The participating farmers of the Kameoka Carbon Minus Project make bamboo biochar with a smokeless carbonizing ring kiln and apply bamboo biochar to the vegetable garden soil (McGeevy and Shibata 2010). By using a simple carbonizing device, individual small-scale farmers can make biochar easily; in urban areas, however, the carbonizing device must not emit smoke or smell.

When developing urban organic agriculture using biochar, it may be realistic to establish a carbonization kiln or furnace that converts a massive amount of biomass into biochar in the suburban or rural areas and distribute the biochar to individual producers. Then, the question about what fertilizer sources can be found to combine with biochar and whether any stakeholder provides the required materials needs to be addressed. Depending on the type and scale of farm management, biochar and organic fertilizer production needs to be examined.

Introduction of biogas digester reduces environmental problems of livestock-raising farmers and improves their livelihood through promoting organic farming. Likewise, introduction of carbonization devices for making biochar will contribute to improving soil, cultivating organically, and raising livestock by farmers. As with biogas digester, the carbonization devices range from expensive to cheap models; they are selected by the users depending on construction materials and required functions. Analyzing and examining such feasibility is required for developing recycle-based urban organic agriculture in tropical developing regions.

Box 10.1 Utilization of Compost and Biochar from Siam Weed (*Chromolaena odorata*) **in Cambodia**

Siam weed (*Chromolaena odorata*) is a tropical shrub of the Asteraceae family widely distributed in upland fields and fallow lands in tropical regions. It is listed in "100 of the World's Worst Invasive Alien Species" by IUCN (International Union for Conservation of Nature) and is often known as a weed harmful to agricultural crops. But in Cambodia, the weed is mixed with upland weeds such as tropical whiteweed (*Ageratum conyzoides*) and livestock manure to be composted and then used. Farmers empirically know that these plants can be used as green manure and compost materials. Our team made biochar from Siam weed and applied it for tomato cultivation (Fig. 10.6). As a result, biochar from Siam weed was found to contain more phosphorus (1917 mg/kg) than rice husk biochar (483 mg/kg) so that the biochar could be more a useful soil amendment for crop cultivation (Lorn et al. 2017). Biomass available near farmers may possibly be used as biochar by the farmers.

Fig. 10.6 Siam weed and Siam weed biochar in Cambodia

10.2 Organic Agricultural System in Agatho Corporation, Indonesia

10.2.1 Opening and History of the BSB Farm

The BSB (Bina Sarana Bakti) Foundation was established in 1984 in Bogor, Indonesia, based on the thoughts of Fr. Agatho Elsener, OFMCap. This initiative is fully supported by the Socio-economic Development Commission of the Indonesian Church Guardian Conference, especially Fr. Gregorius Utomo, Pr, and Ibu C. Djoeariah, SH. The Ministry of Environment of Indonesia also offered recommendations on the establishment of this foundation in 1985.

At first, the foundation was expected to be a development information center because at that time Father Agatho thought that the development in Indonesia was going in a reversed direction. However, because the theme of development was considered broad and less clear, agricultural development was chosen to move especially toward organic farming. Father Agatho was very inspired by a book entitled *The One-Straw Revolution* written by Masanobu Fukuoka. The main theme explained in the book is that "nature has worked as it should and humans only support it." This thinking underlies the making of organic agriculture as a means of establishing the BSB Foundation (Fig. 10.7). Trials of various natural farming practices have been carried out by Father Agatho.

The financial crisis that occurred in Indonesia in July 1997 had a serious adverse impact on the agricultural practice; prices rose by 300% for pesticide and more than 200% for seeds and chemical fertilizers. As a result, farmers could no longer bear the expenses of buying the materials so that they stopped farming. However, this condition has no effect on the BSB Foundation because all needed materials such as compost and seeds are produced independently on-site. Farmers began to raise awareness of switching to organic farming using cheaper manure. After 1997, there was a surge of visitors and guests who came to visit the organic farming to benefit from the comparative studies. The nongovernmental organizations (NGOs) began to be seriously

Fig. 10.7 Building and farm in the Bina Sarana Bakti (BSB) Foundation

involved in assisting organic farmers; and what is encouraging is the change in farmers' attitude toward appreciating compost, manure, and nature. NGOs prepare to establish BSB as a place of educating and training farmers on organic farming. Through the efforts of PT Agatho Agro, vegetable produce began to be exported to Singapore in 2001. PT Agatho Agro and partners also asked the Australian agency to certify the organic agriculture at BSB in order to expand the export of their produce. Export of vegetables to Singapore continued until 2004. After then, the BSB focused on meeting Indonesian domestic demand of green produce. This was made possible because of the increasing awareness of Indonesian people about healthy food, which in the end also becomes a lifestyle of Indonesian people.

10.2.1.1 Basic Concept and Profile of the Bina Sarana Bakti (BSB) Foundation

Since 1987, all land owned by BSB has been used for organic farming. Agricultural practices follow natural law and totally stop using all forms of synthetic chemical inputs including pesticides and fertilizers. This practice is known as one of the pioneering developments of organic agriculture in Indonesia.

The BSB Foundation seeks to understand the evolution of nature and the reality of human development. It turned out that the two paths are very different; they are almost analogous to the Creator and the creation. According to the Creator, everything is there to serve others, whereas human beings think everything are all for their own use.

All elements in nature support each other, hold for the common good, and even give themselves for the survival of others. They believe that this organic attitude is the basis for sustainability. In nature, everything produces more than its own need and gives the extra to the needed others; thus, nature is sufficient for all. To build the future, selfishness must be overpowered by the organic attitude. The vision of BSB is "Living in Harmony with Others, Nature, and God (Creator)"; and the mission is preparing and developing various means so that every human being can devote himself to serving one another, nature, and God (Creator).

10.2.1.2 Management in the Bina Sarana Bakti Foundation

BSB is an "organic development center" that focuses on developing agricultural cultivation and fostering the organic attitude. Father Agatho argued that the failure in agriculture was not caused by the inability to apply organic technology, but more due to attitudes and mentality. Therefore, one of the important aspects is education and training.

BSB has five separate production gardens; they are Asti Garden (ASTI), Merak Garden, Rumah Bawah Garden (RB), Mendawai, and Organic Parks. ASTI is located around the dormitory that is situated in an area of approximately 1 ha used for dormitories, offices, and production land. The dormitory has 60 beds in total and 49 active beds. The harvest is mostly used to meet the needs of the dormitory kitchen. Merak Garden is located 1 km from ASTI and has a land area of about 0.25 ha consisting of 200 beds in total and 174 active beds. This is an educational and training production area with high production results. The Rumah Bawah Garden has approximately 13 ha in area divided into several blocks. The garden consists of nursery areas, PT offices, marketing, composting sites, seed stocks, organic gardens, vegetable shops, and markets. Kebun Mendawai that is dedicated to producing seeds has approximately 250 beds and screened house buildings. Kebun Taman Organic Parks have a focus on ecotourism and education, which consists of several permaculture stations used to present sustainable human settlements through ecology and design. Agroforestry stations present an artificial resource by explaining what the functions of agroforestry are. Herbal stations display and provide information and knowledge about medical plants and traditional medicine. Cultivation/nursery stations present the process of producing vegetables from seeds to mature crops ready to be harvested. Animal farming stations present farm demonstration plots to raise various types of domestic animals that are processed organically so that visitors can learn organic ways and techniques of maintaining and utilizing animals.

10.2.2 Advantages of Farm Location in Bogor and the Valley

BSB is located at the foot of Mount Pangrango, which has an altitude of 800–920 meters above sea level with a land slope of around 3–5%; the land area is 15 hectares. The farms are located in tropical regions that tend to be wet with the rainy season lasting from November to May and the dry season from June to October. The average temperature is 20.5–22 °C, and the average rainfall reaches 160 mm per month with an average humidity of 75–88%. The land is safe for vegetable cultivation because the erosion rate or the possibility of landslides is low. Based on these conditions, farmers focus on vegetable and horticultural crops; the crop plants cultivated consist of tubers, leaf plants, vegetable fruit plants, and legume crops. Besides these plants, other herbal plants to supplement the biodiversity include turmeric, ginger, galingale, rosemary, fennel, oregano, lemon balm, mint, and other legume crops (Table 10.1).

10.2.3 Technologies and System to Produce Organic Agricultural Products

10.2.3.1 Activities of Developing Organic Agriculture

Organic farming developed at BSB is a holistic cultivation system to improve and develop the health of agroecosystems including biodiversity, biological cycles, and soil biological activities. The cultivation activities emphasize the application of

Table 10.1 List of commodities planted in BSB farm

Root	Legume	Vegetable	Fruit
Carrot	Red bean	Spinach	Tomato
Sweet potato	Pea	Water spinach	Cherry
Cassava	Peanut	Caisim	Oyong
Beet root	Long bean	Pakcoy	Eggplant
Radish	Snaps	Cauliflower	Cucumber
		Cabbage	Green chili
		Broccoli	Red chili
		Curly lettuce	Corn
		Red lettuce	
		Siomak lettuce	
		Head lettuce	
		Endive	
		Basil	
		Basillus	

management practices that prioritize the use of inputs from cultivation activities on the land based on the consideration of adaptation to environmental conditions. This cultivation system is achieved by the implementation of biological and mechanical methods without using synthetic chemicals to meet all demands within the system. The efforts in developing organic farming are as follows:

(a) Developing overall biodiversity in the system
(b) Increasing soil biological activity
(c) Maintaining soil fertility in the long term
(d) Recycling waste from plants and animals to restore nutrients to the soil so as to minimize the use of nonrenewable resources
(e) Increasing the use of land, water, and air properly and minimizing all forms of pollution resulting from agricultural activities
(f) Dealing with agricultural products by emphasizing good practices at all stages to maintain organic integrity and product quality

When developing organic farming, the map of the area and the layout of the planned gardens must be first sketched out before the construction starts. The tasks include the brick and garden area, location and direction of the terrace, location and size of beds, as well as the selection of plants, seed calculation, organic fertilizer, vegetable pesticides, costs, and amount of labor, among others. Proper locations of water tanks and nursery sites as well as layout of nurseries must be taken into account so that the crop and gardening needs during the production period can be effectively managed (Fig. 10.8).

The planting beds are made permanently to ensure proper management of soil fertility and plants. Mini elephant grass (*Pennisetum purpureum* Schumach) are planted around the planting bed to prevent soil washing by surface flow due to rainwater and excessive surface flow and to strengthen the planting beds to prevent soil

Fig. 10.8 Terrace planting

erosion or landslide. The beds are oriented westward so that the plants growing in the beds get maximum sunlight for photosynthesis. Dimensions of standard beds are 1 m wide × 10 m long × 20–30 cm high with 300 cm distance between two adjacent beds. This bed arrangement is intended to prevent land exploitation due to unplanned use, minimize erosion caused by the cultivation process, create plant diversity, and maintain soil fertility on every productive land. Further, this arrangement makes it easier to overcome the problem of plant-disturbing organisms in case an imbalance occurs at any time and to facilitate farm analysis to calculate the required inputs.

10.2.3.2 Mixed Cropping

The cultivation system used to achieve organic vegetable cultivation implements intercropping system (mixed cropping) and crop rotation, which aims to increase the soil productivity and break the cycle of pests and diseases. The intercropping system is a system of planting two or more types of plants in one bed at a time with respect to the family and the nature of individual plants and the combination of plant properties such as repellant nature, age of harvest, sun requirements, water requirements, nutrient requirements, root growth, and canopy size. The purpose of intercropping systems is to maximize the productivity of a field, control pests, and maintain soil fertility. Each combination is selected according to the needs to cultivate the land and adjusted according to the nature of the plants.

Several types of combinations that have been carried out at BSB include as follows:

(a) Improved fallow (combination with green fertilizer): Improved fallow of the primary plants and green fertilizer plants is practiced. The latter including *Sesbania, Tephrosia*, and others that are planted in the same bed as the primary plants contribute nutrients to land. The use of green fertilizer in combination with primary plants is intended to make barren land or infertile soil fertile.

(b) Relay cropping (combination of plants in sequence or relay): Relay cropping is used for plants that have either long life or short life by planting long-living and short-living plants together. When short-living plants are harvested, the long-living plants are still growing in the same bed. The available space can then be used to plant short-living plants again so that one bed can produce one crop of long-living plant and two or more crops of short-living plants.

(c) Repellent (combination of insect-repellent plants, OPT (*organisme pengganggu tanaman*)): This type of combination connects the primary plants with the repellant plants that have a strong odor such as onions, basil, lemongrass, rosemary, and others. The objective is to keep pests away from the primary plants so that they are not damaged by pests.

(d) Companion: This type of combination is the planting of two or three different types of mutually beneficial plants simultaneously in one bed (Fig. 10.9).

10.2.3.3 Crop Combination and Rotation

Crop rotation is arranged in such a way that each bed is in a certain plot with the same cycle of crop rotation. Crops of legumes and green manure are planted alternately or inserted with the primary vegetable plants. The types of legumes include *Crotalaria juncea* and *Crotalaria mucoides* which are grown for around 3 months; then the green leaves are cut periodically to be composted or immersed in fresh form into the plot. Each plot will get the same treatment in one cycle of crop rotation.

The purpose of crop rotation is to break the breeding of pests so that there is no explosion of pests or diseases while maintaining soil fertility. There are three types of crop rotation practiced at BSB:

Fig. 10.9 In-house vegetable cultivation

(a) Short rotation pattern: Short rotation pattern is done by planting legume plants or other plants for the first time and then replacing or rotating the legume plant with other types of legume plants.
(b) Medium rotation pattern: Medium rotation pattern is done by planting legume plants followed by fruit crops such as corn, tomatoes, or chili during the next season. Afterward, root crops (tubers) such as beets, carrots, and sweet potatoes are planted during the third season. The rotation pattern can also be carried out by planting legumes; then leaf crops such as pakcoy, spinach, or lettuce during the next season; and root crops such as beets, carrots, or sweet potatoes during the third season.
(c) Long rotation patterns: Long rotation pattern is done by planting legumes and then leaf crops such as pakcoy, spinach, or lettuce during the next season; fruit crop such as corn, tomatoes, or chili is planted during the third season; and root crops (tubers) such as beets, carrots, or sweet potatoes are planted during the fourth season.

Several factors concerning the principle of combination and rotation plantings are considered (Fig. 10.10):

(1) Friendly or opponent plants: Friendly plants include mutually supportive plants such as carrots and scallions, peas and lettuce, cabbage and tomatoes, as well as corn and beans.
(2) Plants with specific natures and nutrient uptakes: The pattern of root development and the age of harvest should be considered by combining shallow-rooted plants with deep-rooted ones such as corn with caisim.
(3) Plant habitus: Plants having upward growth with many branches must be combined with shade-resistant plants such as celery and kale.
(4) Season: During rainy seasons, the spacing regulation is a priority when taking into account the need for light and more pathogen populations caused by high moisture. During dry season, however, the vegetable growth is more optimal because of the available optimal light intensity and relatively few pest diseases.
(5) Diversity of surrounding plants: A variety of plants are grown to avoid pest infections.

Fig. 10.10 The practice of plant protection by removing the infected plants manually

10.2.4   Vertical Nutrient Recycling System of Root, Mulch, and Plant Residue in Soil

10.2.4.1   Mulch

The agricultural land used for cultivation of organic vegetables in the BSB generally has a slope of 3–5%. Therefore, adequate soil conservation techniques are needed to maintain soil stability and water sustainability in the long run. Mechanical soil conservation techniques such as making bench terraces with beds following the soil contour or cutting slopes or perpendicular sides can be combined with vegetative techniques by planting grass or other plants to reinforce terraces. All these efforts are carried out so that long-term soil fertility can be sustained. Grass planted around beds is needed to strengthen the beds and prevent soil erosion and loss of nutrients. The erosion control is of importance especially for sloped areas with terracing systems without overexploiting available water sources.

Another effort to control erosion is implemented by mulching during both dry and rainy seasons. Using mulch is the most effective erosion prevention technique; mulch will reduce water evaporation from the soil surface to maintain soil moisture. Mulching materials come from organic matter, and hence, mulch also becomes a source of soil organic matter to maintain soil organic matter. Besides, mulch also acts as a soil stabilizer to protect the soil surface from blows of raindrops directly. Further, mulch also plays an important role in controlling soil temperature to alleviate heat loss from the soil. Organic materials that can be used as mulch come from crop residues and prunings from the hedgerow system provided that they are not infected by plant diseases.

10.2.4.2   Water and Nutrient Management

Water management that is an important consideration in the cultivation of organic farming includes efforts to provide irrigation water for plants (irrigation) and to drain excess water not used by plants (drainage). The irrigation water used for vegetable crops is derived from springs located in BSB's garden as well as from rainwater storage ponds (Fig. 10.11).

Nutrient management to improve soil productivity and maintain optimal soil biological activity is carried out by rotating crops and utilizing crop residues, manure, legume plants, and green manure. Because nutrients contained in soil are lost by harvesting the crop, fertilization is necessary to maintain, preserve, and improve the soil fertility and health in the long run.

Organic matter plays an important role and is a key factor to the success of organic crop systems and maintaining soil productivity and quality. Organic materials supply nutrients directly and indirectly so that they play an important role in improving soil physical, chemical, and biological properties that cannot be separated from the aspects of soil fertility. One effective way to increase soil fertility is

Fig. 10.11 Plant
protection and watering

by applying compost from animal manure and crop residues. The process of composting organic matter is summarized as follows:

(1) Manure is poured at the place provided (special fertilizer room).
(2) On top of the manure, old leaves or grass are piled up from the ground up with a thickness of about 10 cm above the manure pile.
(3) A thin layer of dolomite is spread evenly to the top of the pile of leaves or grass.
(4) Steps 1–3 are repeated until the overall pile reaches a full height of 1.5 m.
(5) The pile is flushed with water to maintain moisture and to accelerate decay.
(6) The pile is allowed to mature, and it will be ready in about 3 months.

10.2.4.3   Cropping Management

The organic management of vegetable farming takes into account soil fertility and the entire ecosystem to produce species diversity to suppress pests, weeds, and diseases while maintaining soil health and reducing nutrient losses. The diversity of crop production is produced through the following:

(a) Applying a rotation pattern of vegetable crops including legumes and others. The diversity of vegetable plants is maintained directly in the garden (not monoculture), while the diversity of cereal plants is carried out in a crop rotation system. Vegetable plants are planted in beds measuring 1 m × 10 m × 30 cm, but the bed size can be adjusted to adapt to field conditions to facilitate soil processing and plant growing. Vegetable planting is done by growing more than two types of plants per bed based on age of harvest, root growth and shape, size of plant canopy, and growth needs including sunlight, water, and nutrients.

(b) Selecting different types of vegetable plants or planting cover crops to cover the soil surface to the possible maximum extent is practiced. On sloped land, the surface of the soil must always be covered with plants.

(c) Choosing a planting method or planting design is carried out to harvest vegetables in stages according to market needs. Planting plans are carried out appropriately so that sustainable vegetable production is maintained.

(d) Planting insect-repellent plants (OPT) is implemented to control natural pests and reduce the consumption of vegetable pesticides such as tephrosia, lavender, lemongrass, and kenikir by planting basil around beds and pest trap plants such as gamal and mimba around the garden as well.

10.2.4.4 Pest Control

Organic farmland has a high biodiversity with rare occurrence of pests or diseases because the following pest control methods are implemented:

(a) Holistic control: Holistic control is carried out for pest control starting from garden planning such as arranging the direction and location of beds to planting hedges that prevent the spread of pests and diseases through wind and insects. Garden planning is important and must be done appropriately in order to create a good environment for all living things.

(b) Preventive control: Preventive control aims to prevent the occurrence of OPT (animal, insects, or weeds) attacks. It is carried out by planting the crop according to the season in order to understand the risks and types of pest infections, implementing appropriate land treatment, paying attention to the types of plants, and making combinations and crop rotation such as the short-rotation red spinach plant with a red bean plant.

(c) Curative control: Curative control is done by removing the infected plants manually or indirectly by using vegetable control to break the pest cycle. If the holistic control is working, the preventive approach will not be needed.

10.2.5 Wisdom of Success for Managing Organic Agriculture

The organic products at BSB are sold directly to consumers through their distributors spread out across Jakarta, Depok, and Bekasi. This direct-sales system cuts through the long supplying chain that has been continued in conventional marketing systems. Both farmers and consumers benefit from this direct-sales system in terms of crop prices and profits. Efforts are made to keep loyal consumers by periodically educating the consumers on "the principles of trust."

Fig. 10.12 Packaging and labeling of product

To meet the demands of consumers, partnerships and collaborations are conducted with farmers around Bogor. BSB invites the farmers to participate in organic farming. The requirement to become BSB's partner is to have the same vision and mission toward organic farming and follow all organic farming procedures and techniques. In order to maintain the consumer trust in partner farmers' products, regular monitoring and assistance are conducted for partner farmers.

The ability of BSB employees has been enhanced by education and training on organic cultivation. To disseminate this organic agriculture, BSB regularly organizes organic farming training sessions and workshops, which are attended by various groups such as students, farmers, practitioners, and other professionals to transfer knowledge from the experts to the participants.

To inspire the sustainable development of organic agriculture in Indonesia, BSB also consistently collaborates with universities to conduct academic research and assessment on organic farming. One of the collaborations is to make BSB organic garden a place of learning and practicing fieldwork and internships for students, as well as a place to conduct research for researchers and academics. The visitors who want to know about the organic farming process can visit the farm to be spiritually active in the BSB garden. In addition to seeking fresh vegetable products, BSB is currently developing processed organic products from fresh organic produce and participating actively in the organic farming movement through the Indonesia Organic Alliance (Fig. 10.12).

10.3   Biogas Plant and Local Organic Agriculture in Ogawamachi, Japan

10.3.1   Role of NPO Foodo Ogawa and Local Currency

The biogas plant in Ogawamachi is operated by NPO Foodo Ogawa. This organization was founded in 2002 to enhance the continuous development of regional food industries and the local society by recycling locally derived resources such as renewable materials and energy. Mamoru Kuwabara, the organization's representative director, has experiences as engineer in overseas operation commissioned by the Japan International Cooperation Agency (JICA). Kuwabara claimed that he was impressed by the small-size biogas facilities used abroad and has been working toward building such facilities in Ogawamachi. Half of the 8 million yen needed for this endeavor was paid for by 105 investors, and the remainder was financed by a loan from a general incorporated association "AP Bank" that was founded by musician Takeshi Kobayashi. Construction of the biogas plant was launched in 2008. As shown in Fig. 10.13, expenses of the biogas plant are subsidized by Ogawamachi as commission expenses of the district's domestic garbage processing and other related business activities. The bank loan is to be paid off in 10 years.

NPO Foodo Ogawa built an experimental plant to collect research data on methane fermentation and slurry digestion for establishing the current biogas plants. The slurry is digested in the methane fermentation process into farming fertilizer for growing rice and vegetables; technical data of the quantity and quality of the digested slurry (effluent) and its fertilizing effect on plant growth were collected. In addition, the organization has tried to set up the standards and methods of using the digested slurry. The information was disseminated to farmers who receive the digested slurry as fertilizer.

NPO Foodo Ogawa has continued to provide programs that actively involve local residents engaged in the business of garbage recycling. For example, trainees from foreign countries are educated by local residents on the substance cycles involving the biogas plant and organic agriculture. Further, the organization has led the region in holding symposiums for introducing advanced technologies of biogas plants to residents. For example, the organization has conducted research in collaboration with Honda R&D Co., Ltd. on small-scale cogeneration of biogas and has promoted

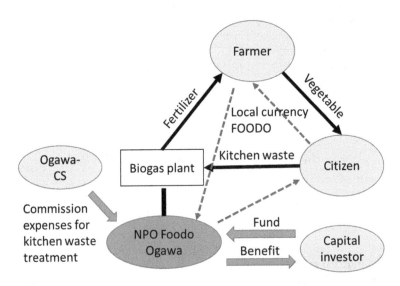

Fig. 10.13 Partnership among concerned parties

the development of stable fermentation technologies in collaboration with Tokyo University of Agriculture and Technology (TUAT) by inviting TUAT researchers to monitor the fermentation progress of the biogas plant (Kobayashi 2016).

The garbage discarded by local residents is an important component of biogas plant operation. Using appropriately sorted garbage, the biogas plant can be operated normally to produce high-quality liquid fertilizer that can be applied by farmers to grow quality produce. NPO Foodo Ogawa issues a local currency known as FOODO to mediate the bartering activities among those involved. This currency is to be distributed as a coupon by the operator of the biogas plant to credit local residents who provide garbage. Residents can exchange these coupons for vegetables from the farmers who use the digested slurry. In turn, farmers can exchange the coupons for more digested slurry from the biogas plant. The capital fund of this coupon is provided by the net cost savings of 20 yen/kg (excluding operation cost) to dispose of garbage through reusing instead of burning in Ogawamachi. Coupons equivalent to 3000 yen for 150 kg of garbage or 20 yen/kg are distributed to families to promote garbage recycling in the region, local producers, and consumers of the agricultural products.

10.3.2 Biogas Plant Structure

As shown in Fig. 10.14, Ogawamachi Biogas Plant (OBP) has a unique structure in which two horizontal fermentation tanks are connected in series. Organic matters are fermented into organic acids in the first tank, and then the acid is fermented into methane gas with the simultaneous production of digested slurries in the second tank. The digested slurry fluid can be used as liquid organic fertilizer. The fermented liquid of 0.5 m^3 per day overflows from the first 12 m^3 fermenter tank to the second 8 m^3 fermenter tank. The organic acids produced during methane fermentation in the first tank are discharged to the second fermentation process to be digested so that levels of nitrogen, phosphoric acid, and potassium contained in the digested slurry should be sufficiently high. Concentrations of organic acids contained in the fermented slurry from the first and second fermenter tanks of the OBP analyzed by TUAT laboratory are shown in Table 10.2. In the first fermenter tank, several organic acids with high concentrations are detected, whereas in the effluent of the second tank, only small amounts of acetic acid and propionic acid are detected. The ion concentrations recorded in the methane fermentation are shown in Table 10.3. Nitrogen is detected in the form of ammonium instead of nitrate. Concentrations of nitrogen, phosphoric acid, and potassium in the effluent from the second tank are higher than those from the first tank. The ammonium nitrogen in the effluent of the second fermenter tank is approximately 0.3% that is slightly higher than that in the liquid fertilizer used as foliar spray. Another analysis indicated that the heavy metal concentrations in the digested slurry are extremely low that meet the criteria stipulated in the fertilizer control regulations in Japan. In addition, enteropathogenic *Escherichia coli* was not detected in the slurry. Raw biowastes are well known to

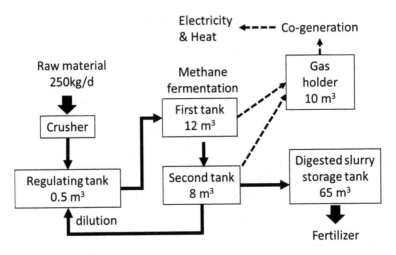

Fig. 10.14 Flowchart of materials in the biogas plant

Table 10.2 Concentrations of organic acids in methane fermentation effluent

Properties and organic acid	Unit	First tank	Second tank
pH	–	5.95	7.51
VFAs	g/L	40.81	0.34
ORP	mV	−122	−181
Acetic acid	mg/L	18551	251
Propionic acid	mg/L	21409	110
Isobutyric acid	mg/L	3831	0
n-Butyric acid	mg/L	1400	0
Isovaleric acid	mg/L	733	0
n-Valeric acid	mg/L	1542	0

Table 10.3 Ion concentrations in methane fermentation effluents

Ion	Unit	First tank	Second tank
Na^+	mg/L	864	1249
NH_4^+	mg/L	2413	3792
K^+	mg/L	1585	1898
Ca^{2+}	mg/L	34	26
Mg^{2+}	mg/L	16	21
PO_4^{3-}	mg/L	384	1339
Cl^-	mg/L	1096	1596
NO_2^-	mg/L	nd	nd
NO_3^-	mg/L	nd	nd
SO_4^{2-}	mg/L	nd	nd

nd: not detected

contain pathogenic bacteria that originate from tissues of diseased animals and people as well as from feces, urine, and exudates of healthy carriers (Sahlstroom 2003).
However, the digested slurry of OBP contains no pathogenic bacteria because garbage wastes from residential kitchen and school cafeterias are used as the raw material in the processing plant.

Each day, 250 kg of garbage is collected from 300 residential houses and the
school cafeterias. After being crushed, the garbage is diluted with 0.5 m³ circulated
liquid from the second fermenter tank and then used as the input to the OBP process. The hydraulic retention time (HRT) is 24 days for the first fermentation tank
and 16 days for the second fermentation tank. The emerging biogas is temporarily
stored in a 10 m³ gasbag; the digested slurry discharged from the second fermenter
tank is stored in a 65 m³ liquid storage tank for 100 days. The biogas is used in 1 kW
small-sized cogeneration to produce electricity and hot water. The hot water is sent
to the heating pipes placed at the bottom of the fermenter to heat up the digester
content.

10.3.3 Use of Methane Fermentation Digested Slurry

As mentioned in the previous sections, an important objective of this biogas plant is
to produce digested slurry and apply it to the farming fields as organic fertilizer.
NPO Foodo Ogawa published the guide on the rate of digested slurry to be applied
for replacing chemical fertilizer as follows: 30 t/ha as a basal fertilizer and 10 t/ha
as an additional fertilizer applied in the paddy fields. In the paddy fields without
compost, the digested slurry with nitrogen 1.5- to twofold of the nitrogen contained
in standard chemical fertilizer leads to the same yield as the chemical fertilizer. In
the case of wheat, addition of the digested slurry results in significantly more yield
than the conventional fowl droppings. In the case of broccoli, use of digested slurry
to the plug seedlings before they are planted increases the root weight by twofold
and promotes the growth of the planted seedlings. In other cases, the data show that
potherb mustard and lettuce grow relatively faster as a result of applying digested
slurry. Although conveying the digested slurry to the fields takes some effort, the
slurry liquid is convenient to apply in the field because it needs no further dilution.

Furukawa and Hasegawa (2006) conducted experiments with spinach and mustard spinach plants to confirm the fertilizing effects and the safety of the digested
slurry. Their results indicated a nitrogen absorption rate of 22 g N m⁻². The nitrogen
content in the leaves was 1.47 g kg⁻¹, which is similar to that contained in the leaves
of the plants grown using chemical fertilizer and cattle compost fertilizer. The fertilizing effects were similar to the immediate effects of chemical fertilizer, and the risk
of disease caused by pathogenic microorganisms was low.

The farmers participating in this system pump the digested slurry from the liquid
storage tank of the biogas plant into 500 L tanks mounted on a pickup truck and then
transport the slurry to the field. In the paddy fields, the digested slurry is mixed with
the irrigation water prior to being applied to the crops. The paddy fields supplied

with digested slurry are not watered for the first 2 days. The digested slurry is drained from the tank, mounted on a pickup truck, and sprayed to the field through a specially designed pipe distribution system. Adding wood vinegar to the digested slurry in the tank neutralizes slurry pH in addition to promoting crop intake of nutrients and suppressing the vaporization of ammonium.

Box 10.2 Application of Digested Slurry from Methane Fermentation to Rice Cultivation

Anaerobic digestion for biogas production leads to several changes in the composition of the resulting digested slurry (digestate) as compared with the original feedstock; the slurry composition is relevant to the availability of macro- and micronutrients for plant growth after the slurry is applied in the field (Möller and Müller 2012). Especially, the effect of applying the digestate on rice production in the paddy fields was not well recognized. This experiment was carried out to demonstrate the effect of the digestate from Yagimachi biogas plant in Kyoto Prefecture applied to the paddy fields as liquid fertilizer (Miho et al., 2004). The rice known as "Hitomebore" was grown in the paddy fields in Fuchushi, Tokyo, under three experimental conditions: no fertilizer (NF), chemical fertilizer (CF), and the digestate (DS). A 48 m^2 plot was set up for two repetitions of each study. In the CF plot, basal fertilizer consisting of 40 kg/ha of nitrogen, phosphate acid, and potassium each and additional 40 kg/ha of nitrogen, phosphate acid, and potassium each was applied. In the DS plot, the same quantity of nitrogen as in the CF plot is applied. In both plots, basal fertilizer was applied after transplanting was completed (Fig. 10.15), and additional fertilizer was applied just before ear emergence. The cultivation results shown in Table 10.4 indicate that the DS plot has slightly higher yield of brown rice than the CF plot. Further, as the yield components are considered, the thousand kernel weight from the DS plot is greater than that from the CF plot. This indicates that the DS plot has more total yield than CF. However, there is no significant difference noted between the yields of the plots using the digested slurry and the chemical fertilizer. The protein content that is strongly related to the taste of brown rice was lower in the rice produced in the DS plot. Lower protein content gives the expectation of good taste.

Another issue is the odor caused by using the digestate as fertilizer; it is caused by ammonium nitrogen contained in the digestate as an important nutrient. The odor of ammonium is unpleasant for people who live around the field where the digestate is widely applied. Results of detailed measurement of applying the digestate to the fields are shown in Figs. 10.16 and 10.17. For wheat cultivation, the ammonia is emitted to the air from the field where the digestate is applied without flood water. Immediately after spreading the digestate to the soil at 30 °C, the maximum ammonia emission rate was 705 mg m^{-2} h^{-1}. However, the emission decreased to 54 mg m^{-2} h^{-1} after the

(coninued)

Box 10.2 (continued)

ammonia had been absorbed into the soil. During the next 2 days, the ammonia emission rate decreased exponentially and became almost undetectable. Conversely, for the flooded field, the maximum ammonia emission rate at 30 °C was 20 mg m^{-2} h^{-1}, which is 1/35 of the maximum ammonia emission rate from the non-flooded field. Although the ammonia emission rate was low, it persisted for about 5 days. This creates a difficult challenge when using the digestate for rice cultivation, particularly in urban areas where residential houses are close to the paddy fields. The odor caused by emission of volatile ammonia from the digestate can be alleviated by lowering the digestate pH to decrease the vaporization of ammonium component.

Fig. 10.15 Field setups of measuring gas emission from the paddy field plot after digestate application. The emitted gas is collected and temporarily stored in a small chamber on the paddy field for a certain time

Table 10.4 Yield of rice by using digested slurry as liquid fertilizer

Plot condition	Stubble (stubble/m^2)	Grain (grain/stubble)	Ripening rate (%)	Grain mass (mg/grain)	Yield (t/ha)	Protein (%)
No fertilizer	22.6	667	87.7	21.84	2.87	8.2
Chemical fertilizer	22.9	837	87.6	21.85	3.60	9.1
Digested slurry	22.5	818	86.6	21.90	3.91	8.9

(coninued)

Box 10.2 (continued)

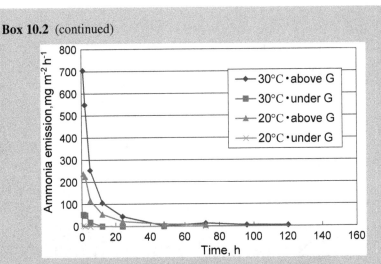

Fig. 10.16 Ammonia emission from paddy field not flooded with water after applying digested slurry. "Above G" means that the digested slurry was applied above ground surface; "under G" means that the slurry was applied into soil of the paddy field

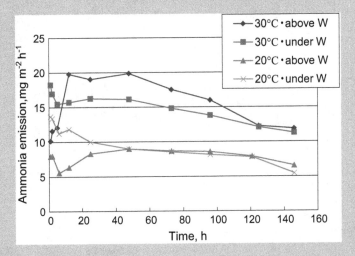

Fig. 10.17 Ammonia emission from flooded paddy field after applying digested slurry. "Above W" means that the digested slurry was applied above water surface; "under W" means that the slurry was applied into water of the paddy field

10.3.4 Local Organic Agriculture in Ogawamachi

According to Oguchi (2012), Ogawamachi is well known for its organic agriculture, and the reputation is attributed mostly to an outstanding farmer, Yoshinori Kaneko. Growing up in Ogawamachi and having graduated from the prefectural farmer's academy, Kaneko succeeded his father to practice agriculture. Since 1971, Kaneko has been working in "Shimosato Farm" to grow organic produce. In 1975, he developed a ten-member membership-only subsistence farming system with membership fee paid monthly. This project was unsuccessful, and in 1977, he switched the project to a system known as the 'Thanks System" that is based on a collaborative local production and local consumption concept. Since 1981, he has begun to sell the bagged vegetables and sets of eggs. This approach to sell agricultural produce is currently extended to 40 families, direct-sales depots, restaurants, morning markets, and local food industries. Orito (2014) analyzed this farming system as follows: The "Thanks System" was traditionally known in this region as a gift exchange system. Although this system did not require monthly payments, consumers who received vegetables paid gratuities willingly. Because agriculture depends on the climate, the farmer in general cannot earn enough income to sustain the farming. This system supports farmer's efforts by supplementing their income at a certain level. Therefore, the concept comes from the thought that the agricultural products are viewed not as commodities but as food that sustains human health and livelihood.

Among the 397 commercial farmers in Ogawamachi, 33 engage in organic agriculture. This high number of organic farmers is related to the acceptance of trainees working on Shimosato Farm, which began in 1979. Currently, four farmers accept trainees in a long-term program for 1–3 years. Among these trainees, some become farmers in Ogawamachi after the training period, and these new farmers lead groups to engage in organic farming. The trainees learn agricultural techniques and management practice, and create several links with the local residents by participating in local events and village works. The farmers who accept trainees also become a bridge between the new farmers and the local residents to develop sales channels of the agricultural products and support machine rental.

Shimosato Farm pioneered several sales channels by cooperating with local food-related businesses and industries. The cooperative entities include sake brewers, noodle makers, soy sauce producers, and tofu shops. For example, rice is provided to the corporation for commercial production of natural sake at a price of 600 yen per kg, and soy is sold to tofu shops at 500 yen per kg. In both cases, the prices double those if the products were sold to non-industrial consumers. Such a system of developing new channels to ensure sales of agricultural products invigorates the regional economy and enables continuous development of organic agriculture.

According to Iizuka (2009), women, who play a significant role in organic agriculture in Ogawamachi, comprise a large segment of new farmers. These new female farmers, 20–40 in age, have lived in cities; they become farmers to experience the agricultural living. They share a crucial viewpoint against pollution in cities and the use of agricultural chemicals as well. Hence, they are attracted to engage in organic agriculture for enabling the recovery of natural products to enhance healthy lifestyles.

References

BAJ: Bridge Asia Japan (2012) Biogas digester promotion project for small scale farmers in Hue City, Vietnam. Japan Fund for Global Environment, FY 2011 Activity Reports (in Japanese). https://www.erca.go.jp/jfge/subsidy/organization/act_repo/report23/055.html. Accessed 10 May 2019

BAJ: Bridge Asia Japan (2013) Biogas digester promotion project for small scale farmers in Hue City, Vietnam. Japan Fund for Global Environment, FY 2012 Activity Reports (in Japanese). https://www.erca.go.jp/jfge/subsidy/organization/act_repo/report24/pdf/04_057.pdf. Accessed 10 May 2019

BAJ: Bridge Asia Japan (2014) Biogas digester promotion project for small scale farmers in Hue City, Vietnam. Japan Fund for Global Environment, FY 2013 Activity Reports (in Japanese). https://www.erca.go.jp/jfge/subsidy/organization/act_repo/report25/pdf/04_085.pdf. Accessed 10 May 2019

BAJ: Bridge Asia Japan (2017) The project for technology training of organic vegetable cultivation and strengthening of farmer organization in Hue, Vietnam. Asian Co-operative Cooperation Fund, Activity Report 2017, p. 24 (in Japanese). http://ccij.jp/book/pdf/etc20170614_01_01.pdf. Accessed 10 May 2019

BAJ: Bridge Asia Japan (2018) The project for technology training of organic vegetable cultivation and strengthening of farmer organization in Hue, Vietnam. Asian Co-operative Cooperation Fund, Activity Report 2018, p. 19 (in Japanese). http://ccij.jp/book/pdf/etc20180611_01_01.pdf. Accessed 10 May 2019

BAJ: Bridge Asia Japan (n.d.) Bridge Asia Japan Website (in Japanese). https://www.baj-npo.org/. Accessed 10 May 2019

Furukawa Y, Hasegawa H (2006) Response of Spinach and Komatsuna to biogas effluent made from source-separated kitchen garbage. J Environ Qual 35(5):1939–1947

Glaser B, Lehman J, Zech W (2002) Ameliorating physical and chemical properties of highly weathered soils in the tropics with charcoal – a review. Biol Fertil Soils 35:219–230. https://doi.org/10.1007/s00374-002-0466-4

Hoi An Tourism (n.d.) Tra Que Vegetable Village. http://hoian-tourism.com/the-heritage/hoi_an_traditional_occupations/tra-que-vegetable-village. Accessed 10 May 2019

Iizuka R (2009) The entry process into agriculture and the view of life of organic farming woman – a case study on Ogawa-machi. J Rural Life Soc Japan 52(2):12–21

Kobayashi K (2016) Development of monitoring methane fermentation using three dimensional fluorescent analysis. MS thesis, Tokyo University of Agriculture and technology (in Japanese)

Lehmann J, Joseph S (eds) (2009) Biochar for environmental management: science and technology. Earthscan, London

Lehmann J, Joseph S (eds) (2015) Biochar for environmental management: science, technology and implementation, 2nd edn. London/New York, Routledge

Lorn V, Tanaka H, Bellingrath-Kimura SD, Oikawa Y (2017) The effects of biochar from rice husk and Chromolaena odorata on soil properties and tomato growth in Cambodia. Trop Agric Dev 61(3):99–106. https://doi.org/10.11248/jsta.61.99

McGreevy SR, Shibata A (2010) A rural revitalization scheme in Japan utilizing biochar and eco-branding: the carbon minus project, Kameoka City. Annals Environ Sci 4:11–22. http://hdl.handle.net/2047/d20000347

Miho Y, Tojo S, Watanabe K (2004) Utilization in cultivation and its environmental load of digested slurry from biogas plant. J Jpn Soc Agric Machinery 66(3):77–83. (in Japanese)

Möller K, Müller T (2012) Effects of anaerobic digestion on digestate nutrient availability and crop growth: a review. Eng Life Sci 12(3):242–257

Nguyen VCN (2011) Small-scale anaerobic digesters in Vietnam: development and challenges. J Vietnam Environ 1(1):12–18. https://openjournals.neu.edu/aes/journal/article/download/v4art2/v4p11-22/

Oguchi K (2012) Establishment of the regional relationship and the development of organic agriculture: a case study of Saitama Prefecture Hiki Country Ogawa Town. J Rural Stud 18(2):36–43

Oikawa Y, Phan QD, Phan VBD, Nguyen VL, Yamada M, Hayashidani H, Tanaka H, Tarao M, Katsura K (2018) Charcoal application farming with livestock for small scale farmers in Central Viet Nam. In: Agroecology for food security and nutrition, proceedings of the international symposium on agroecology in China FAO, Rome, pp 197–210. http://www.fao.org/3/CA0153EN/ca0153en.pdf

Orito E (2014) Teikei as "morotomo" relationship embedded in agrarian rationality: the experience of the "Orei-sei" system at Shimosato farm. J Environ Sociol 20:133–148

Sahlström L (2003) A review of survival of pathogenic bacteria in organic waste used in biogas plants. Bioresour Technol 87(2):161–166

Steiner C, Teixeira WG, Lehmann J, Nehls T, JLVd M, Blum WEH, Zech W (2007) Long term effects of manure, charcoal and mineral fertilization on crop production and fertility on a highly weathered Central Amazonian upland soil. Plant Soil 291:275–290. https://doi.org/10.1007/s11104-007-9193-9

Yamada M, Kawabata Y, Oikawa Y (2014) Research, education and extension of environmental technologies in developing countries: case study of Tokyo University of Agriculture and Technology. J Arid Land Stud 23(4):185–191. http://nodaiweb.university.jp/desert/pdf_APCSEET2013/185-191_Yamada%20et%20al%20rev.pdf

Yamada R (2008) Diagnosis, design and evaluation of diversified farming in the Mekong Delta of Vietnam: based on the farming systems research approach. Norin-tokei-kyokai, Tokyo (in Japanese)

Index

© Springer Nature Singapore Pte Ltd. 2020
S. Tojo (ed.), *Recycle Based Organic Agriculture in a City*,
https://doi.org/10.1007/978-981-32-9872-9